Marketing for Scientists

MARKETING FOR SCIENTISTS

How to Shine in Tough Times

Marc J. Kuchner

Washington | Covelo | London

Island Press is a trademark of The Center for Resource Economics.

Library of Congress Cataloging-in-Publication Data

Kuchner, Marc J., 1972-
   Marketing for scientists : how to shine in tough times / Marc J. Kuchner.
      p. cm.
   Includes bibliographical references and index.
   ISBN-13: 978-1-59726-994-0 (pbk. : acid-free paper)
   ISBN-10: 1-59726-994-8 (pbk. : acid-free paper)   1. Science--Vocational
guidance. I. Title.
   Q147.K83 2012
   502.3--dc23

                                                        2011026553

Printed on recycled, acid-free paper (logo)

Manufactured in the United States of America
10 9 8 7 6 5 4 3 2 1

Keywords: Island Press, marketing, scientist, science and media, science
blogs, academic tenure, jobs for scientists, funding proposals, promotional
savvy, branding, scientific conferences

Contents

Acknowledgments

First, I'd like to thank Kelly O'Brien for teaching the class that helped inspire this book, and for helping me develop the concept and conduct the first interviews. I'd like to thank my science mentors: Mike Brown, Wes Traub, David Spergel, Jennifer Wiseman, Bill Oegerle, and in fond memory, Myron Lecar. I'd like to thank my music-business mentors: Jan Buckingham, Roy Elkins, Bob Dellaposta, Rex Benson, Martha Irwin, and Dusty Wakeman. And I'd like to thank my parents, Joan and Eugene Kuchner, for encouraging my interest in science in the first place.

This book is partly based on a Facebook group called "Marketing for Scientists," where many scientists contributed ideas and musings. I'd like to thank all the members of the group and especially Stella Kafka, Amber Straughn, Sara Seager, Amir Give'on, Peter Ong Lim, and Zoë Fonseca-Kelly for being my co-admins, helping launch the group, and helping keep the conversation going. Here are some of the hundreds of scientists and other experts who contributed to the book via this group: Sandy Antunes, Daniel Pendick, Ruslan Belikov, Nicholas Suntzeff, Malcolm Fridlund, Luisa Rebull, Charley Noecker, Amy Lo, Heidi Hammel, Adam Burgasser, Angela Speck, Agnes Kim, Dan Vergano, and Michael Lemonick.

Another crucial source of input for the book was a series of workshops. I owe thanks to the many people who encouraged me to run them, helped out, and gave me feedback through them. Thanks to Evgenya Shkolnik, Kevin Fisher, Marc Postman, Ori Fox, Mario Perez, Dennis Bodewitts, Kelle Cruz, and Stuart Vogel for the invitations. Special thanks to Stella Kafka and Amil Patel for getting up in front of everybody and helping me lead the workshops.

Many colleagues read drafts of this book and sent me comments and suggestions. Thank you Jon Gardner, Mario Perez, Cole Miller, Rita Sambruna, Rachel Anderson, Paolo Tozzi, Fergal Mullally, Andrew Youdin, Mark Shelhamer, Pamela Millar, Lucy McFadden, Amy Simon-Miller, Sarah Stewart-Mukhopadhyay, Thayne Currie, Aki Roberge, Chris Stark,

Hannah Jang-Condell, Ruslan Belikov, Sally Heap, Laura Giza, David Leisawitz, Phyllis Bernstein, Bill Danchi, Malcolm Neidner, Ken Carpenter, Chris Mooney, Chris Lintott, and Derek Sivers.

Other colleagues and friends contributed by participating in interviews and conversations, formal and informal. Thank you Jeremy Kasdin, Anand Sivaramakrishnan, Michael Liu, Rich Townsend, Subhanjoy Mohanty, Paul Kalas, Andrea Dupree, Shri Kulkarni, Steve Kilston, Jim Austin, John Debes, Anne Kinney, Mike Cresswell, David Leckrone, Harold Marshall, David Seiler, Robert Naeye, Frank Reddy, Jennifer Ouellette, Shai Oster, George Musser, John Marburger III, Nell Greenfield-Boyce, David Wilcove, Dennis Overbye, John Mather, Joshua Wurman, Randall Larsen, Jay Morris, Robert Thompson, Cristina Eisenberg, Erika Nesvold, Markos Georganopoulos, Dana Berry, Deborah Leckband, Pensri Ho, Irene Klotz, Katherine McAlpine, Will Barras, Genevive Bjorn, David Goldhaber, Mark Kushner, David Pinsky, Dovie Wylie, Joan Kuchner, Rita Colwell, Ludmilla Kolokolova, Bill Reach, and Robert Walker.

I want to thank my agent, Judy Heiblum, for believing in this strange idea for a book. I want to thank Peter Yezukevich for contributing his formidable graphic design talents. Then there's everyone at Island Press who worked with me and helped teach me about the book business: Erin Johnson, David Miller, Jaime Jennings, Maureen Gately, Mike Fleming, and many others. I'm especially grateful to my editor, Barbara Dean, for lending her logic and grace to these pages, swimming across the ocean with me three or four times, and for talking to me as though I were an actual writer. And most of all, I would like to thank Jennifer Nuzzo for patience and love, for marrying me and protecting me, and for being my first follower in most of my crazy schemes. I couldn't have done this without you.

Marc J. Kuchner
June 16, 2011

Introduction

In 1995, the Endangered Species Act was in deep political trouble. The House of Representatives was considering a bill that would cut sixteeen million dollars from the endangered species activities of the Fish and Wildlife Service and completely abolish the National Biological Service. The speaker of the house, Newt Gingrich, was saying that it made little sense to spend money on species protection because extinction is just "the way life is."[1] A group of concerned scientists arranged to meet with Gingrich, wishing they could mitigate this potential disaster by reasoning with him about the value of these endangered plants and animals. We scientists often like to think that logic and a solid argument ought to carry the day.

If only it were that simple. If only reason alone sufficed to change people's minds and spread the good word about science. And if only, decades later, scientists weren't still suffering from lack of support for their causes and for their research.

On the contrary, times seem to have gotten harder for science in the United States since 1995. University endowments in the U.S. plunged an average of 18 percent in 2009, leading to widespread furloughing and the shuttering of whole departments.[2] The House of Representatives passed flat budgets for the major federal science-funding agencies in 2011.[3] The economic pressures our country is facing extend to the lives of graduate students and postdoctoral fellows, who are facing dim job prospects.[4]

Coinciding with these financial challenges are cultural ones. Many newspapers and magazines have dropped their science sections or replaced them with lean "news-you-can-use" coverage. Anti-science sentiments

seem to be mainstream nowadays; members of Congress are asking the
public to decide which National Science Foundation (NSF) grants to cut.[5]
In a recent survey, only 40 percent of U.S. adults agreed that "human beings, as we know them, developed from earlier species of animals."[6]

And never mind trying to save our country from the current tide of
ignorance—simply trying to forge a career in science has long been a mystifying endeavor. Rejection rates for papers submitted to some scientific
journals are traditionally 70 to 80 percent. NSF funding rates for new in-
vestigators are typically less than 15 percent. For decades, applicants for
tenure-track positions have often numbered in the hundreds. How are we
meant to cope with these odds?

Like many scientists, I've long worried about these problems. But now
I've found a tool that I think can help us deal with most of them, maybe
all of them. This tool can help us succeed in science or in academia, or
launch an alternative career. It can help us find jobs, win grants, attract
students, get tenure, communicate to the public, or promote science to
Congress. It can give us the perspective we need in order to adapt ourselves to difficult times.

Tom Eisner, a Cornell professor of ecology who was one of the scientists meeting with Newt Gingrich that day in 1995 had this tool in mind
as he walked past the security guards into Gingrich's office. David Wilcove, Professor of Public Affairs and Ecology and Evolutionary Biology
at Princeton's Woodrow Wilson School, told me the story:

> Tom Eisner brought to the meeting a small vial that had a little
> cutting from an endangered mint plant in Florida that Eisner felt
> might harbor some interesting [compounds that could be used as]
> new insecticides and pesticides, and presented it to Gingrich. Gin-
> grich was intrigued and asked if he could keep the vial! Then, ul-
> timately, Gingrich told the scientists that he would make sure that
> he would not let the ESA be gutted.
>
> —David Wilcove

Instead of walking into Gingrich's office looking to argue, Eisner
walked in eager to start a conversation. His prop—the mint plant cutting—
broke the ice and gave him an opening that led to two more meetings

with Gingrich. During those meetings, he forged a new relationship that in turn helped shape U.S. ecological policy for the better.

Gingrich made good on his promise. He gave a speech where he embraced the "values and goals" of the environmental movement. Later on, when a bill that would have undermined the ESA cleared the house resources committee, Gingrich did not let the bill come to the floor for a vote.[7]

The heart of Eisner's approach—starting a conversation, building a relationship—might be called many things: advocacy, education, or salesmanship. Whatever you call it, it seems to me that scientists of every kind need to learn more about it. The word I prefer to use may make some of us uncomfortable, but now is not the time to be squeamish. That word is marketing.

What is this?

The square pattern at the beginning of each chapter is a mobile barcode. If you have a smartphone, you can scan in the barcodes and they will take you to webpages with links and media related to the chapters in this book. These resources may also be accessed by going directly to www.marketingforscientists.com.

CHAPTER ONE

Business

I am an astrophysicist, an expert on the theory and observations of planetary systems around other stars. I earned my PhD from Caltech in 2000 and then went to the Harvard-Smithsonian Center for Astrophysics on a prize postdoctoral fellowship to study planet hunting and planet formation. I won a second fellowship in 2003, the Hubble postdoctoral fellowship, and went to Princeton to work on new methods for finding Earthlike extrasolar planets. Now I work at NASA Goddard Space Flight Center as a staff scientist. Like my fellow mid-career scientists, I write and referee papers, advise students and postdocs, serve on panels, organize conferences, write proposals, and give many, many PowerPoint talks.

As I was forging my scientific career, I was never quite sure if I was doing it right. I watched some of my colleagues succeed in science, some even launch huge programs, while I saw others fail, and I wondered why the chips fell as they did. Some of my scientific colleagues mentioned "marketing" every now and then, but there was no official recognition that this craft might have a role in how we developed our careers and spread our scientific ideas. Some colleagues would even roll their eyes or speak in a whisper when they said the word.

With help from some wonderful mentors, I am lucky to have landed a good permanent job in a field that many people pursue as a hobby. But it was through my hobby that I first began to learn systematically about marketing. On the weekends and in the shower, I write songs—country songs. I aim to write more or less the kind of contemporary country song you hear on today's country radio. And as confusing as I found the astrophysics business, I found the songwriting business even more mystifying.

When I started out writing songs, I focused only on my songwriting craft, trying to make my lyrics sound ever more honest and compelling, my melodies more catchy and original. I read many books and magazines and websites about country music. I learned about lyrics, song forms, and guitar technique. I started taking regular trips to Nashville to record the songs I was writing.

Eventually, the songs started sounding really good to me. Really, really good. I was sure one of them would be a hit. But the responses I got on my first trips to Nashville were disappointing. Everyone kept telling me: great songs, man, but the music business is a business! I nodded and smiled, though I didn't know quite what that meant.

Well, I found out what it meant when I began earnestly trying to sell the songs I was writing. I mailed out dozens of packages of CDs, but one after another came back unopened. I called people up to ask if they liked my songs, and they hung up on me.

This rejection hurt. It hurt for two reasons: it hurt because my songs remained unsold, and it hurt because of the way I was rejected. The cozy collegial climate of academia that I was used to doesn't exist in the music business, or in most businesses for that matter. In science, when your paper is rejected or your proposal is turned down, you get a polite letter explaining the decision and offering feedback. Maybe this feedback from our academic colleagues is often late or incomplete. But in business, if the customers don't want your merchandise they walk straight out of the store and they don't fill out the comment card. If they really don't like what you are selling, they sometimes even go online and leave a trail of nasty comments about your work that you and everyone else can read.

Fortunately, I kept on writing songs, trying to find someone who would listen, and asking experts for advice. Then a turning point came for me when I decided not just to study the craft of songwriting, but to try to understand the music business as a business. I filled my bookshelf with books about marketing and started trying to apply what I read to attracting the attention of artist managers and music publishers. I studied sales, and learned how to build working relationships with people I didn't know, starting from scratch. I sent out a newsletter with handy tips. I wrote articles in trade journals like Nashville's *Music Row Magazine*. I worked on developing a coherent brand.

At some point my efforts started paying off. I started getting calls from managers of country artists and bands who liked my songs. I was so giddy that at first I let them have the songs for free. Then that changed; I started earning royalties that paid for trips to Nashville and new music gear.

I haven't exactly become an A-list honky-tonk hero, nor do I expect to quit working as a full-time scientist, a job I love. But I have now had almost twenty songs published by music publishers in Nashville, Los Angeles, and in Europe, and I have garnered about the same number of "cuts": recordings of my songs by country artists. Four of these cuts have received radio airplay. Last year one of my songs was chosen as the best demo of the year by *Music Connection Magazine*. This year, part of another one of my songs appeared in the show *Making the Band*, produced by P. Diddy; it aired on MTV, MTV2, and BET.

Anyway, one day I had a conversation with a postdoc who seemed to be writing the same paper over and over again, complaining that nobody was citing his work, nobody seemed to be listening. The frustration was familiar. It struck me that scientists and other academics are often in the same position I was as a beginning songwriter: writing papers nobody reads—like songs nobody hears.

We write paper after paper, and cast these papers out into the ether, hoping they will land on fertile ground, and often receive back nothing more than a slow drip of dutiful citations from a few close collaborators. We spend months crafting long proposals and preparing for job applications, only to face rejection after rejection with little explanation. And worst of all, every scientist seems to have some kind of painful story to tell about doing groundbreaking work that mostly got ignored, then watching someone else put out a press release on the topic and hog all the credit for the idea.

One might be tempted to think that the many slights and rejections we scientists must suffer are somehow a necessary part of our education. But I don't think that way anymore. My experience with the music business has taught me to cherish every bit of feedback I can get, and not to think of the hundreds of unreturned phone calls or ignored pitches I must face as signs of personal failings. It was this change of perspective, and the pressure it removed from my life, that first made me want to try systematically applying what I learned in the music business to the world of science.

So I set out to put together the marketing toolkit that I wish I had had when I was starting my science career (and my music career). I began with the hodgepodge of wisdom I'd learned in the music business. Then I interviewed a series of scientists in various fields at different career stages to find out how they marketed themselves. I began interviewing other professionals; I talked to press officers, reporters, and staff at funding agencies, as well as experts in outreach, government, and science policy. As I gained confidence, I started giving workshops on marketing to help coach the students, postdocs, and junior scientists at my institution, and to gauge their needs and responses. Finally, I started a Facebook group called "Marketing for Scientists," invited my colleagues to join, and discussed my ideas with the members of the group. I collected many new concepts and anecdotes from this group, and also abandoned many dead-end ideas thanks to the feedback I received.

The result is this book. It's organized a bit like a science textbook, because—I can't help it—that structure appeals to me. First, I talk about the theory of marketing; then I talk about applications. I go through some of the fundamentals of sales, relationship building, and branding, and try to concoct a kind of science-marketing perspective. Then I go back and examine some familiar institutions of science such as papers, talks, and press releases through this perspective.

In the process of writing the book, I've come to understand that marketing already threads its way through the fabric of today's scientific and academic institutions. As Princeton ecology professor David Wilcove told me, "Even scientists who don't think they are marketing their work are marketing. The introduction to a technical paper is a piece of marketing. When you write a grant proposal, you're basically marketing." As Caltech astronomy professor—and former chair of my department—Shri Kulkarni puts it, "Being a good scientist is half science and half marketing." If you are a scientist, then, you may find yourself already acquainted with some of this material, though perhaps in a different form.

As an (astro)physicist, I like to start by writing down the fundamental theorems. So I'm going to start with the first thing I learned about marketing back when I was writing songs that nobody listened to—a kind of basic underlying principle. You can skip ahead if you like; there's stuff later on that's more fun.

In any case, whether you consider yourself a marketing skeptic or a marketing fanatic, there is a picture of marketing I want to paint for you, one that I think is still new to most scientists. It's a sensibility that has emerged in the Internet era: marketing based on genuine two-way conversations and community-building. I think scientists can all use this new approach without fear of being selfish or disingenuous—to improve society, conquer ignorance, and share our passion for discovery.

Now, more than ever, marketing is for scientists.

CHAPTER TWO

The Fundamental Theorem
of Marketing

 There's a simple idea that I came to realize is the foundation of business and marketing, something like the Schrödinger equation of quantum mechanics. It looked innocuous and slightly alien to me when I first uncovered it. Here it is.

Everything you get from other people comes because you met someone else's needs or desires.

I like to think of this statement as a kind of fundamental theorem. It has no imaginary numbers or Greek letters. But the rest of marketing is largely ramifications of this one hypothesis: you don't get anything for free.

Now, the fundamental theorem applies to everybody. So every deal between two people has a certain kind of symmetry to it. Consider this: you write a song for use in a TV show. You give the TV producer permission to use the song. He gives you some money. The two of you have participated in the "markets" for songs and money. You met his needs, and also he met your needs, just like the fundamental theorem says. That's how the music business works—how any business works.

But how, you may ask, can the fundamental theorem of marketing apply to the transactions of a scientist? For example, say you applied to ten graduate schools. You were admitted to three of them. That's all. There's no money changing hands, and no apparent product for sale.

But you had a desire; you *wanted* to go to graduate school. According to the fundamental theorem, someone else apparently *wanted* you to show up at graduate school, too. You met someone's desires and they met yours. That's the fundamental theorem in action.

You might say that the process of being admitted to graduate school is symmetric, just like the process of selling a song. You received something (the admission) and you also gave something (your application and your demonstation of confidence in and preference for the school). You might say that these things you exchanged had more or less equal value. Someone in the administrative structure of the universities you applied to gets to smile a big smile, or maybe sigh a sigh of relief when he sees the applications arrive in the mail. Or maybe it's not one person with a big smile—maybe it's twelve people in the department office, the grants office, and the dean's office with little smiles or little sighs of relief.

I doubt that most students think about both sides of the equation while they are writing their graduate school applications. But the thing is, no matter what taboos or diversions might be there to distract you from it, every step in your career somehow follows the fundamental theorem. Every one! Even in science, as some people say: you scratch my back, I'll scratch yours.

I struggled for a long time with this notion; it did not sit well with me at first or even seem useful for a scientist to think about. But then, once you begin thinking harder about the fundamental theorem, you start to see the world differently. You start to turn the theorem around. I want to get things from other people—*how can I fulfill other people's needs and desires?*

For that matter, you start to wonder what *are* other people's needs and desires? Sometimes it's easy to figure out what's expected of you; the graduate school application form says that you should write your name on the top of each page. But sometimes it's hard to figure out what other people want. For example, what exactly should you write in your application essay? Sometimes people go to great lengths to conceal what pleases them and what motivates them. It seems to me that the graduate school application process, for example, cleverly conceals the motivations of the people behind it.

Marketing: A Definition

This discussion brings us to the notion of marketing. Here is my preferred definition of the word:

> Marketing is the craft of seeing things from other people's perspectives, understanding their wants and needs, and finding ways to meet them.

In other words, if you want to get into graduate school, you have to peer past the formality of the applications forms and see through the cordiality of the admissions office. You have to understand who wants you to apply and why. You have to figure out what they need and what you can offer them. Maybe you succeeded in this game; while you were writing those application essays you paused for a moment to consider the needs of the real humans who would eventually read them. That was marketing, right there.

As I read the literature, I found many definitions of the word *marketing*. Some of them talked about identifying products and deciding on their prices—concepts that are tricky to transfer to science (though we'll try to in a little bit). But as Harvard Business School professor of marketing Theodore C. Levitt said, marketing invariably views "the entire business process as consisting of a tightly integrated effort to discover, create, arouse, and satisfy customer needs."[1] That seems to me to be the bottom line: thinking about the needs of other people.

Thinking about the needs of other people turns out to be pretty profound. As I learned more about marketing I was shocked over and over again by how much effort it takes to do it right. Some people may find it intuitive, but I was impressed over and over by how much of marketing I found bewilderingly nonintuitive—at least from my scientist's perspective.

What's Hard about Marketing

I took a mind-blowing marketing class from Kelly O'Brien, owner of TurningPointe, a consulting firm focused on new media and management strategies.[2] She was tall and fashionably dressed, and she worked with name tags and flip charts. She threw little brightly colored plastic toys on the tables to set the mood. When I walked in on the first day, before the aforementioned mind blowing, I felt momentarily like I was in a scene from the evil business world, or maybe from the television sitcom *The Office*.

Kelly began the first class with a deceptively simple exercise that helped me understand the surprisingly subtle magic of seeing the world through other people's eyes. Let's say you owned a movie theater—a giant multiplex in a suburban area. What are your daily concerns and worries? "Take a moment and brainstorm with me," she said, and began jotting down ideas from the students on one of those loathsome flip charts.

Concerns of a Movie-Theater Owner

- Is the popcorn stocked?
- Do we have new reels to show next month?
- Is our ad in the paper up to date?
- Who's working the ticket booth tonight?

Now Kelly said, "Let's pretend you'd like to see the latest James Bond movie. What are your needs and concerns?" A few moments of discomfort and a few squeaks of her magic marker later, she had written down these ideas.

Concerns of Someone Who Wants to See a Movie

- What time is it playing?
- Where is it playing?
- Does that theater have a big screen and good sound?
- Should I just wait and see it on DVD?

"What I wanted to show," I remember her saying, "is that thinking like the movie-theater owner feels very different than thinking like a moviegoer. The moviegoer does not know the theater owner or the staff, or care who's working tonight. The moviegoer takes popcorn for granted. But when you stand in someone else's shoes, everything is different."

This might sound strange, but when I switch from thinking about the movie-theater owner to thinking about the moviegoer, I briefly feel a faint physical sensation. Maybe it is some subtle change in blood flow in my head or the release of some hormone.[3] I get a similar sensation walking into a room full of new people, or walking onstage to perform in front of an audience. Switching perspectives like this requires you to draw on your knowledge of what it might be like to be someone else—knowledge we often don't have much of. You have no choice but to wing it. And I find it faintly physically uncomfortable.

But imagining that we are someone else is something we do all the time. It's even part of our science education. Consider this exercise in perspective changing.

Concerns of a Scholar Writing a Paper
- Did I calculate the error bars right?
- Are the figures in the right place?
- Will my coauthors sign off on this?
- Can I get this out the door before I get scooped?

Concerns of Scholar Reading a Paper
- Did these authors remember to acknowledge and cite my work?
- Does this paper support my hypotheses or agree with my data?
- Can I use these findings in my next project?
- Are there any figures in here I can use in my next talk or proposal?

Perhaps you had a conversation with your advisor in graduate school where you learned to take your readers into account when you write a paper—to carefully choose what figures to show, what references to cite, and what conclusions to highlight. I would like to point out that as you took this journey of trying to think like your customer, you were *marketing*.

A Philosophical Comment

I mentioned that I struggled with the fundamental theorem at first. Well, I'm not the only one to find this notion challenging. Several scientists I've spoken to about marketing have told me that there was no way they could reconcile the fundamental theorem with their worldviews. Are people really that selfish that they only give something away when there is something in it for them? Am I really that selfish? Are you? These questions stopped me in my tracks for a bit.

I had a discussion about the fundamental theorem with Rus Belikov, a crisply logical staff scientist at NASA Ames Research Center (and my former postdoc). He summed up the fundamental theorem by saying, "In other words, there's no such thing as altruism, approximately." I had to agree; the statements are logically equivalent, though maybe Rus's version sounds a little bit colder.

But consider this: taking the fundamental theorem to heart pushes you to think about the needs, desires, yearnings, and dreams of your fellow human beings. There's nothing cold or selfish about that. Here's a little chart that seems to sum up our options.

> Recognize the fundamental theorem. ❯ Work on helping your fellow human beings.
>
> Or
>
> Ignore the fundamental theorem. ❯ Maybe try to help other people or maybe not, depending on your innate level of altruism.

To me, this chart shows the fundamental theorem to be a beautiful thing whether or not it is always true and no matter what it may imply about human nature. It's beautiful simply because it compels you to think about the needs of other people. After looking at it this way, I've decided that it's a good principle to follow. If you are still uncertain, I challenge you to take a moment and think about people in your life, *their* needs and how you can help them. It's kind of like trying to think of a good birthday present to give someone, something that that person would really like. Unless you really have no heart, giving a gift and making someone happy feels darn good.

★　★　★

What Do People Want?

So the fundamental theorem compels us to think hard—using our full rational powers—about what people want and need. What do people need and want? Well, in the process of developing this book and giving workshops on marketing at colleges and other scientific institutions, I've put this question to the attendees, often a room full of postdocs: What do people want? The responses I've gotten have been revealing.

There's always a momentary silence, while the room of PhD scientists adjusts to the shock of being asked such a trivial-sounding question. Then usually, someone breaks the silence with a suggestion like this: "They want to feel appreciated." Or: "They want a sense of purpose." Or: "They want a sense of beauty and wonder." That might well be true. But mentioning these particular desires first strikes me as a little bit backward.

Let me offer you a picture of human desires that gets the priorities mostly right. Abraham Maslow, in a 1943 paper entitled "A Theory of Human Motivation," created a handy chart of fundamental human desires, his famous "Hierarchy of Needs." Maybe you ran into this chart in high school or college. I have reproduced it here in Figure 2-1, flipped around so the most urgent needs are on top.

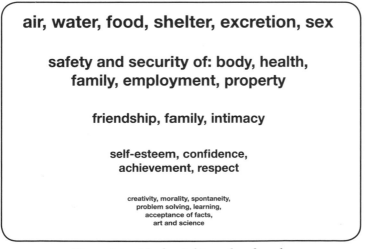

Figure 2-1   Maslow's hierarchy of needs.
Source: Pete Yezukevich.

Maslow's hierarchy is a prioritized list; our needs on the top must be met before we can think about what's further down. First we need to be able to breathe. Then we need a bite to eat, and probably a bathroom. Then, if we have the right stimulus, we may be thinking about sex. Only when those most urgent animal needs are met can we start thinking about whether the roof is sturdy. Then, if the house is safe and sound, and there's food in the fridge, and our backs don't ache too much, we can start to worry about whether we have any friends to hang out with. Once we know we have people who support us and care about us, we can begin to wonder about the stars and the origin of mitochondria and the symmetry of buckyballs (and how to market them).

Yes, we want to feel appreciated and we want a sense of wonder. But first we have to breathe and eat and procreate and enjoy the company of other humans. Indeed, looking at Figure 2-1, and thinking about the fundamental theorem, you might wish you were in a different business. Maybe it would make good sense to pick a business that addresses needs from the top of the hierarchy, like, say, the food business.

Some science marketers have even made this particular leap. Science may be losing some of its perch in our society, but the food culture in the U.S. lately has been blossoming in new ways, even during this recession. Television schedules are packed with cooking shows. The internet buzzes with blogs and articles on food, restaurants, and cooking. And now, at the Atlanta botanical gardens, "food enthusiasts can stop in at a new edible garden, with an outdoor kitchen frequently staffed by guest chefs creating dishes with fresh, healthy ingredients."[4] According to the *New York Times*, "Edible gardens are the fastest-growing trend at botanical gardens, consistently increasing attendance, experts say, along with cooking classes."

But most scientists don't work on anything edible. We have placed ourselves, by choice, at the other end of the hierarchy of needs. So it seems like we will have to work extra hard to help the people around us see that our work is important. That's what the next chapter is about.

CHAPTER THREE

How to Sell Something

 One day, before I became a marketing nut, I was frustrated because my phone calls to Nashville were not paying off. So I went to the bookstore and bought a copy of *Telephone Sales for Dummies*.[1] I remember catching my reflection in the mirror on the way out of the store and thinking: What am I doing? I'm an astrophysicist, not a shoe salesman; I should not need a book on sales, much less one for dummies.

But if you want to be a country songwriter, it quickly becomes clear that you do need to understand sales. Here's how it works. You write a song. Then you send out pitches—maybe hundreds of them. The pitches might take the form of a CD you put in the mail to a record label, or an e-mail you send to a publisher or artist (i.e., a country singer). If you're lucky, you may end up on the phone with the singer's agent or manager, explaining why he or she should pay you money in exchange for a piece of paper, called a license, that gives the singer permission to use the song. That's essentially telephone sales—and it's something I did not do well at first.

Now, *sales* is not strictly *marketing*. It's the craft of directly convincing someone to buy something and then closing the deal. But it overlaps with marketing; it's part of what many people think of when they think of marketing, and it's a skill many scientists have told me they need to have.

Indeed, the other day I had to call up a colleague and try to talk him into putting my data through his data-processing software. After that, I had to call another colleague and convince him to slip me a few thousand dollars from his grant for computer support. Then I had to convince my student that it would be a good idea to get her part of the project done before the conference coming up in February. I realized that sometimes I need to be a salesman when I'm being a scientist just as much as when I am being a songwriter. So allow me to share with you some principles of sales that I think can be helpful to us scientists.

Confidence, Empathy, and Positive Human Interactions

A sale is a kind of intricate dance. Picture yourself as a salesperson, if you will. Imagine if selling a car went like this:

SALESPERSON: Would you like to buy a car?
CUSTOMER: Okay. Here's $40,000.

It sure would be easy being a salesperson if every sale proceeded so directly. But in reality, the closing of a deal looks more like this:

CUSTOMER: Do you think this is the right one to get?
SALESPERSON: That's our most popular model; I'm sure you'll be delighted with it! Which color do you like?
CUSTOMER: The red one.
SALESPERSON: Okay. We can have that one ready to go on Monday. Is Monday okay with you?
CUSTOMER: Sure.
SALESPERSON: Okay. I'll draw up the paperwork.
CUSTOMER: Can I put some of this on my credit card?

In a sale, the actual decision point is an invisible, nebulous thing. In this example, the salesperson never literally asked if the customer wanted to buy the car, and the customer never exactly answered. Instead, the discussion was about colors and delivery dates. But the decision was made and communicated, somewhere between "the red one" and "Can I put some of this on my credit card?" To chase these nebulous customer decisions, the salesperson has to be a master of confidence and empathy, recognizing when the customer wants to move forward by picking up on indirect verbal and nonverbal cues.

You might also say that a salesperson enables people to buy by satisfying one of their needs: the need for positive human interactions. The charm and attention you get from a good salesperson meet a need from somewhere in the middle of Maslow's hierarchy. Having this need met can give us the chance to ponder our other needs, like the need for a flashy red car.

So these are the major themes of sales: providing positive human interactions—reading and giving off verbal and nonverbal cues, and being confident. That's what the following tips are mostly about.

Storytelling

I began the introduction to this book with a short anecdote about Tom Eisner and Newt Gingrich. It's more or less customary to begin a book in this manner—for a good reason. A story can enthrall you and make you feel understood in a way that no amount of straight-up academic prose could have done.

For this reason, telling stories is a crucial sales technique, one you have seen used many times. "I used to eat junk food all the time—when I was bored—when I was lonely. Then, when I joined *Jenny Craig*, my whole life changed. I lost 45 pounds in just a few months. I started fitting into clothes I hadn't worn since high school. All thanks to *Jenny Craig*."

A scientific sales pitch can start with a little story too. "Einstein tried to write down an equation to describe the expansion of the universe. He added a term to the equation that he came to refer to as his 'greatest mistake.' Well, we made these new observations of galaxies that are actually accelerating away from each other—a kind of *cosmic acceleration*. We realized that Einstein's extra term wasn't a mistake at all; this cosmic acceleration corresponds to Einstein's extra term."

But what exactly *is* a story? Maybe it's odd that I need to even ask this question. But until I heard Ira Glass, host of NPR's *This American Life*, explain it, I didn't know the answer myself, not precisely.[2] Here's what I learned. According to Glass, a story is based on two elements:

1. A sequence of events that are causally related. X happened. So Y happened. So Z happened.

2. Moments of reflection when the primary characters pause to di-
gest what's happened so they can change their course of action.

Of course, there is also characterization and setting and dialogue and
whatnot. We'll talk more about other elements of storytelling in a bit,
and later in chapters 6 and 11. But Glass's point is that two ingredients are
the bare minimum you need to make a string of words feel like a story: a
sequence of events, and occasional pauses for reflection.

*I woke up. I looked around the room. I realized that the hawkman was gone,
so I sucked in my breath, picked up my cold gun and ran down the stairs.*

Storytelling is different from the usual language of science. Encyclo-
pedia articles or scientific journal articles, for example, are expositions—
essentially lists of facts. We think carefully about the order in which we
present the material in a paper. But generally, except in a historical over-
view or a discussion of proceedure, scientific papers are largely written
to deemphasize the chronology of events, as though the outcome were
known from the start. Sometimes, if you scrambled the order of the sections and even the order of the sentences, at least the superficial meaning
of a scientific paper would not change.

The opposite is true in a story. Storytelling doesn't require an exposition or list of facts, except to set a scene or introduce a character. And the
sequence of events means everything. In a story, the unfolding events and
the uncertainty about what will happen next (or what has already happened) draw the reader along.

But there is often no reason that a story can't communicate the same
ideas that are contained in an exposition. It might take more words or
more time, and may seem roundabout, or maybe even frivolous. But as I
am sure you intuitively realize, people will remember the story much better. No matter how boring your material is, if you tell it in the form of a
story, it can become captivating.

*The X-rays arrived from the lab. We took one look at them and we could
immediately see that the skeleton had no notches on the femur. The fossil
could not have come from North America! So we packed our tools and
raced off to Cairo.*

Of course, storytelling is not just a way to communicate our experiences or sell our ideas. Stories are the fiber of the human experience. As Irene Klotz, writer for *Discovery News* said to me, "We are all storytellers, every one of us. That's ultimately all we have to do with the time we are here on Earth."

The Stories Your Customers Tell Themselves

James Watson, codiscoverer of the DNA double helix, has been called a master of scientific marketing. A colleague remembers following Watson as he knocked on the door of a wealthy potential donor. "Just as we got to the door, Jim mussed up his hair, bent down and untied his shoelaces, and—looking now like the eccentric scientist—rang on the door."

Ordinarily, we would probably try to make ourselves more presentable before we knocked on someone's door, hoping to make a good impression. Watson's move seems contary to this common sense. In any case, something about Watson's approach must have worked; he secured a large donation.

Marketing guru Seth Godin would say Watson's method was a kind of storytelling. Storytelling is so deeply embedded in human nature and the human experience that it comes up over and over again in marketing. It goes far beyond whatever stories a salesman might literally recite to customers in a sales pitch, for example. Seth Godin likes to say that every customer tells himself or herself his own story about your business.[3] That story, the one that the customer tells to *himself*, is what really sells the product.

Let's talk some more about these stories, the ones customers tell to themselves. Here, Seth Godin is using the word "story" in a slightly different sense than Ira Glass. He's talking less about a sequence of events than about setting and characterization. And he's also invoking a second meaning of the word *story*: a fiction.

I'm pretty sure Watson had in mind a story that he wanted the wealthy man to tell himself: "I am a wise and discerning philanthropist who only supports dedicated scientists, scientists so dedicated to their work they don't even take the time to tie their shoes." Watson may have helped the rich man tell himself this story by providing the right cues. Then the philanthropist participated in his own story by donating money to Watson's project.

Like Watson, companies want their customers to tell themselves a satisfying story about their products. But you can't feed people such stories directly. It only works if the customer writes his own story, starring himself. So instead, companies design products and lay out their stores to appeal to the customer's five senses. For example, when you are buying a car, you slam the door and it makes a solid *thunk*; that speaks of quality construction. You climb inside and smell the new car smell; that makes you think of what a prosperous person you must be that you can afford a new car. Car manufacturers engineer these details, from the new-car smell to the sound of the door slam, to help the customer find the story he wants to tell himself.

In Nashville, there is a similar concept of little sensory details that evoke a whole scene; they make all the difference in a country song as well as in a Toyota lot. In songwriting, we call them "furniture." For example, a song I wrote with Ryan Hydro called "Jack Nicholson" takes place during "Friday midnight poker at the veterans hall." Line two of the song sets the scene: "late-night movie on an old TV set." In this case, a literal piece of furniture, the old TV set, helps set the stage and transport the listener into the song's realm.

Now, this business of encourging your customers to tell themselves stories, which might well be exaggerations, shall we say—well, that makes some scientists uneasy. And it should. We are servants of the truth; we must not encourage people to lie to themselves. Indeed, the story about Watson is probably so memorable because it seems to paint Watson as dishonest, contrary to our visions of ourselves as scientists.

But consider this example. Last weekend I went to a farmer's market and paid six dollars for local, antibiotic-free, sustainably produced free-range chicken eggs. That's a lot of adjectives. That's also three times what ordinary supermarket eggs cost. To justify the purchase, I told myself a story about how I'm a hip young foodie activist, concerned about the environment and animal welfare.

The story I told myself is questionable. It seems more likely that I am just another self-righteous yuppie who's scared of salmonella. But the thing is, the impacts of my purchase and my pleasure in making it are honest and true. It *is* better for the environment and for the chickens if people buy eggs from local, humane farmers. Moreover, I *really did* enjoy the eggs more because of the story I told myself. The label on the package, the cold farmers'-market morning, the photos on Springfield farm's website, and the story I told myself combined to turn the eggs into more than just a refrigerator staple—they became an egg buying and eating adventure.

You will have to judge for yourself how best to utilize this kind of storytelling. But I am happy with my expensive eggs and their story. And I do not want to give back my favorite country songs, just because they are made up. There is a time for us to enjoy the pictures of life we paint for ourselves in our heads, even if they are fictional or rose-colored. Maybe that time comes more often than scientists like to admit. And crucially, we need to recognize that people are telling themselves stories all the time anyway—stories about us and our work—whether or not we pay attention.

Looking Good

With these thoughts in mind about the stories people tell themselves, I examined my professional life and soon found myself wondering: what stories will people tell themselves when they leave my office? Will they tell themselves a tale about how they are smart and important forward-thinking people—the kind of people who need to consult a top scientist working at the cutting edge of knowledge? Or will they spin for themselves a yarn about how they are busy people—too busy to bother with this subordinate wannabe? If I'm not careful, I might end up looking bad in the stories people tell themselves about me.

I started thinking that maybe I should try to engineer the experience my customers have when they interact with me—to provide them with some sensory details and let their minds subconsciously fill in the blanks. First, I decided it was time to clean my desk. Then, I thought about how car manufacturers aim to appeal to all five senses, and so I asked myself, Can I look, smell, and sound like a successful astrophysicist? (I left out taste and feel—I'm not sure how to apply those to astrophysics.) I asked myself: Do astrophysicists wear cologne or perfume? Not much. Do they shower? Yes. Check. Do astrophysicists wear slick designer clothes or ripped concert shirts to work? Not much. Would carrying a nice pen help me look scholarly? Yessiree.

In general, salepeople are taught to pay attention to their appearances and taught to dress well. Making such an effort can probably help scientists as well, and there is even empirical evidence for it. You may be aware of a much-debated 2003 study by Hamermesh and Parker that explored how a professor's appearance can affect student evoluations. The study showed that college professors, male and female, who are "better looking" score significantly higher on these evaluations.[4] Maybe when students see a professor who is attractive, they tell themselves, "I'm not being forced to take a class, I'm just hanging out with my friend, the cool-looking professor."

This survey seems like something to keep in mind if you are a scientist who teaches—and even if you aren't. Your colleagues are all former students; it seems likely that they could be susceptible to the same kinds of influences. The legend about Watson's untied shoes and messy hair may seem to be a counterexample to this notion. I want to point out, however, that Watson was making only superficial adjustments, making himself slightly unkempt, not neglecting his health and physique. In any case, taking care of ourselves and paying attention to our appearance strikes me as one way scientists can ultimately serve the truth better.

Props

There are many conferences and meetings about science careers, and plenty of career advice to be found. One meme you'll often hear at career-development workshops is the importance of having a good "elevator speech." That is, you might be told to practice a two- or three-sentence speech describing your work in simple terms, so in case you meet your congressman or someone else important in the elevator, you can explain what you do in 30 seconds before he or she reaches the eighth floor.

Preparing such a speech is a good exercise. It's a good example of what science-career training has learned from the business world (academics call it "elevator speech"; businesspeople are more likely to call it an "elevator pitch"). It's a device that helps us understand people's tiny attention spans, and the need to be concise and respectful of people's time. A good 30-second speech is a tool we can use in many scenarios, even as part of a much longer presentation. And the process of preparing such a speech forces us to think about the most important messages we have to share with the world, an important marketing exercise.

But I've always found there to be something forced about elevator speeches, and the hypothetical scenario. A canned speech is just a canned speech; it's limited in its power to capture the imagination compared to the more dynamic, interactive, multi-sensory experiences competing for our time these days—via our Xboxes, for example. And according to my informal surveys, real top scientists just don't take time to prepare elevator speeches—except during a career workshop. Maybe this is just because they have become so practiced at it over the years that they can do it off the cuff. Or maybe it's because they realize that when someone asks you about your work, what you say should really be tailored to the person you're talking to—their experiences and preconceptions.

So let's take another look at the elevator speech scenario. What should you do if you meet someone from the Senate Committee on Science, Commerce, and Transportation in the elevator? Marketing gives us an answer, or at least another point of view.

Primarily, marketing is about thinking about other people and their needs. So, first, you should ask the senator what she is working on. Then, if you have something to contribute, take the senator's card, tell her that you think you can help, and ask if it's okay to give her a buzz. Then assemble your thoughts, and the next day when you're back in your office, place that call. But if you don't have something to contribute, just let the poor senator go to work, so that at least you don't make a bad impression.

Now, what if someone asks you about your work? That's a glorious moment: an invitation to promote your science. And it's a moment that the craft of salesmanship tells us how to handle.

I read a story in the *New Yorker* about a guy who goes around Las Vegas selling prosciutto and other specialty foods to fancy restaurants.[5] He has the idea of carrying around an acorn in his pocket to show to his customers. Why? Because the pigs that become the prosciutto eat acorns.

And when you hand your customer an acorn, it evokes an image of happy pigs living in a forest. The little acorn—its weight, its smoothness, its hardness—speaks volumes about that happy scene in the forest that no sales pitch or elevator speech could communicate with words alone. It is a little prop that helps people tell themselves the story they want to hear about pigs that would be wholesome and natural to eat. As we've been talking about, it's that second story, the story people tell themselves, that sells the pork.

One day, Anand Sivaramakrishnan came over from the Space Telescope Science Institute to visit me. Anand works on a new kind of optical technology for astronomical telescopes: an optical mask that blocks some of the light captured by the telescope, effectively turning one big telescope into a constellation of little telescopes and paradoxically increasing its resolving power. When I asked him about his work, he reached into his pocket, pulled out a test version of the optic he was working on, and placed it in my hand. It was an aluminum disk with several circular holes cut in it, seemingly at random (Figure 3-1).

I had heard about Anand's work and I understood the principles of

Figure 3-1   Anand Sivaramakrishnan and his optical mask: an irresistible prop.

this new optical mask in an abstract way. But it was a busy day, and I was not quite in the mood for thinking abstractly about someone else's optical calculations. However, when he placed the mask in my hand, suddenly the way I felt about his concept changed. It went from an abstract concept and a handful of equations to something real. Suddenly I was full of new questions about him and the little prop he had brought.

I wasn't the only person whose attention Anand caught with this prop. Anand and his collaborators came up with the idea to put the mask on the James Webb Space Telescope, a multi-billion-dollar NASA mission designated to replace the Hubble Space Telescope. Unfortunately, the James Webb was already in late stages of design review, and the widespread sentiment was that it was too late to make any changes to it. Nonetheless, Anand began shopping his idea around to the various teams involved in building instruments for the project. Two U.S. teams were too busy to talk to him. So he flew up to Canada where a third team was meeting. This time, he brought with him a copy of the mask, which his collaborator, Peter Tuthill, had cut out of cheap shim stock.

Anand described the scene to me, especially the moment when the Canadian scientists decided to accept his proposal. "The question came up at the science team meeting about how complicated this mask was to make. So I pulled this thing out of my pocket. There was a little bit of silence in the room when people suddenly realized that this would only cost a few hundred bucks, compared to the cost of the whole billion-dollar telescope."

Maybe you don't work on something as concrete as a telescope part. But I bet there is a way to illustrate your work with a prop. Maybe it's a model of a virus or a piece of a drill bit, or even just a vial of red liquid. But props are the magical tools that allow us to start conversations with powerful people. This is how Tom Eisner began convincing Newt Gingrich to support the Endangered Species Act when it was on the chopping block in the House of Representatives, as I described at the beginning of this book. If you want to practice crystallizing and honing your top-level ideas, prepare an elevator speech. But if you want to make a senator stop the elevator and get off at your floor, carry a prop.

Positioning

Back in the 1960s, typical advertising slogans generally focused simply on the qualities of the product ("You'll find the woman's touch in every Purex product"). Sometimes they told how the company was the "best" in some abstract way ("You can't do better than Sears"). Then a rental car company appeared on the scene with new kind of slogan that changed the face of advertising and propelled one of Madison Avenue's most successful ad campaigns.[6]

There weren't many car rental companies at the time, but there was already a recognized brand leader: Hertz. The new car rental company's slogan reacted to this state of affairs. It was risky, maybe even a little goofy, but instantly loveable: "Avis is Only No.2. We Try Harder."

Al Ries and Jack Trout wrote a book about this phenomenon and invented a new term to describe it: *positioning*.[7] By *positioning*, they meant the key quality that sets your product apart from its competitors, explained in a way that addresses the preconceptions of your potential customers. The revolutionary Avis slogan provides this key quality: it explains why people should work with Avis. Never mind that Hertz may have a longer track record and more rental locations. Ries and Trout would say that Avis sucessfully *positioned* itself in the marketplace with its humble slogan, and that this positioning allowed it to flourish. (I love these kinds of stories about the history of advertising. Maybe I've been watching too much of the TV series *Mad Men*.)

Earlier, when I was talking about elevator pitches, I complained that the concept had not been explained to me correctly. Here's what I think might be a better explanation. Folks in business learn that the elevator pitch is not a prepared speech; rather, it's an opportunity to try to understand where your customer is coming from, and help them tell themselves a story about how your product fits in. In other words, an elevator pitch is about the positioning of the product.

For example, I'm an astronomer, and much of my work could be considered esoteric. So I often start my elevator pitch saying, "You know the Hubble Space Telescope?" Of course people know about the Hubble space telescope. Just like they already know about Hertz rental cars. Then I try to tie their knowledge of the Hubble Space Telescope to what I'm working on. For example, I might say that I'm working on a successor to the Hubble Space Telescope, a future telescope that we can use to study planets like the Earth orbiting other stars.

Besides being a way to capture people's attention, positioning is also a display of confidence. In a world where everyone claimed to be number one, it clearly took guts for Avis to admit that they were number two. Customers react well to this kind of confidence; if you're confident about your product, they will be too.

What's In It For Me? (WIIFM)

Here's a related concept from sales that can also help you with your elevator pitches. When a potential customer sees a new business deal for the first time, the customer asks himself, consciously or subconsciously, What's in it for me? This concept is considered so important in sales that it is often abbreviated WIIFM. WIIFM (pronounced "whiff-um") is your first-class ticket into your customer's mind, and a term I'll be using later in this book.

Let's say you are at a meeting and someone hands you a copy of his latest research paper. What goes through your head? You might wonder: Is this something I should pay attention to so that I can maintain my expertise in this field? Does this paper contain a result that affects my current research? Did this paper cite my recent paper?

All of these questions are essentially WIIFM. You might have good intentions and the biggest heart in the world, but you still view the world through your own eyes, and you place your own interests first. It's nothing to be ashamed of. And the people at the meeting whom you hand your papers to will view those pages with their own interests in mind as well.

Customer Service

Forgive me for including this section; it may seem like a dull disquisition on minding your manners. But I feel like at some point in my education as a scientist, I got the message that I was not responsible for thinking about customer service. If, instead, some teacher of mine had explicitly mentioned that this might actually be a component of my science career, I think it might have saved me some grief. As a practicing scientist, I feel like I am often judged by my customer-service skills, even if people don't refer to them as such. If you've ever turned in a referee report late and thereby annoyed your colleagues, you know what I'm talking about.

Customer service is another business concept that can be hard to define precisely and generally. Many of the definitions I've read bleed into other aspects of marketing. But here's a colloquial definition that will do for our purposes:

> Customer service is the craft of helping your customers get what they came for and walk away happy.

I think people have come to expect businesses to play a certain role, the role of a gracious host or maybe even a parent. Sometimes such expectations are justified. When a customer walks into your store, whether it is real or virtual, he enters a strange new environment, and is temporarily unable to take care of himself—a bit like a small child. You can generate goodwill for your business if you recognize this temporary powerlessness and treat your customer as a guest, helping him find his way and enjoy himself, while taking responsibility for whatever issues he can't deal with himself because he is in your domain. Like other aspects of sales, customer service helps meet the basic needs of customers, allowing them the confidence and freedom to contemplate other needs on Maslow's hierarchy.

An obvious example is service in a restaurant. The customers do not know where the silverware is kept. To provide good customer service as a restaurant worker, for instance, a waiter, you must recognize their helpnessness and not expect them to eat with their hands; you must show them where the spoons are or bring some to their table.

The concept extends to businesses that do not interact with their customers in person. For example, here's a story about customer service that they tell new employees at 1-800-Flowers.com.[8] An elderly woman once called 1-800-Flowers.com hoping to have them send flowers to her sister, who was ill. The problem was that the ailing sister lived in a tiny town, far from any florist.

The associate who answered the phone wasn't sure she could help, but she took down as much information as she could about the ailing sister's whereabouts. After making some phone calls, the associate found that this tiny town had a small police station, and she tracked down the lone police officer on duty. She explained the situation to the police officer, who happened to know the sick woman. So the associate arranged for the flowers to be delivered to the police officer, whose home was close enough for the florist to reach. The officer then took the flowers to the sick woman.

The ailing sister was delighted with the flowers and the policeman was delighted with his part in helping deliver them. So, the story goes, 1-800-Flowers.com gained three new customers that day: the elderly woman, her sister, and the policeman—all because of the good customer service the associate provided. The associate recognized that the customer was not an expert in the flower delivery business, so she took responsibility for trying to solve the customer's problem. She helped the customer get what she came for and made her happy by the time the interaction ended. In the process, she generated a wealth of goodwill that extended beyond the original customer.

Now, you can probably think of many ways to provide good customer service, even though, as a scientist, you have no physical storefront and no tangible product to sell. The lowest-level customer-service duties (and often the most difficult for me) have to do with being prompt or on time: answering e-mails and returning phone calls promptly, and arriving at meetings on time (I'm really not good at that). Another, higher, category is about treating customers with care and respect: listening to them patiently, offering assistance, and going the extra mile, as the fabled 1-800-Flowers associate did, without making promises you can't keep. An even higher category is about anticipating your customer's needs, like the waiter who brings you an extra spoon because he anticipates that you'll want to share that four-gallon ice cream cheescake turtle fudge thing you so greedily ordered.

Of course, in this discussion, I have in mind the notion that scientists have customers just like restaurants and teleflorists—a concept I've already been using and I'll continue to try to develop. But when exactly do scientists need to provide good customer service? I'm sure you have plenty of images in your mind of successful scientists who always seem to behave pompously or rebelliously, and professors who don't seem to give a damn.

However, I suspect that if you put most scientists in front of their primary funding organization, the president of their university, the National Academy, or the Nobel Prize Committee, they would instinctively summon some of the helpful resourcefulness of that 1-800-Flowers.com associate. When a scientist is giving you poor customer service, it's probably because you aren't an important client to that scientist. It is up to each of us to determine which people are most important to our scientific careers and to use the craft of customer service to keep them interested in supporting our work.

Calling People by Name

Let's stick with the theme of customer service for a bit. A classic customer-service trick is to learn the customer's name and to use it—often. People love the sound of their own name. For some reason, calling someone by name, rather than saying "hey you" or even "sir," just makes that person like you. It makes you seem confident and familiar.

Somehow, many scientists seem to have forgotten the simple social skill of remembering names and using them. Of course, if you mention someone's name in every sentence, you will sound like a used-car salesman. Maybe some scientists are so afraid of coming off as used-car salesmen that they instinctively overcompensate, and never call people by name. But there is a considerable distance between the used-car salesman and the typical absent-minded scientist, in terms of name usage. I think it's probably safe to say that almost all scientists could stand to increase the rate at which they use names. So I'm going to suggest the following rule of thumb: Use the name of the person you are speaking to whenever you start or end a conversation, and once in every e-mail.

Sure, you're bad with names. So am I. But part of why you are bad with names might be that you aren't accustomed to using them. Remembering names is much easier if you practice saying them; once you get used to saying people's names, I think you will become better at remembering them.

Here's another related magic word that every good salesman knows: *you*. In a fancy restaurant, the waiter doesn't simply ask, "How was the soup?" He is more likely to ask, "How was the soup *for you*?" That's because waiters know that working the word *you* into their sentences earns them higher tips.[9]

Sometimes, It's Good to Interrupt People

Like many scientists, I hold a weekly meeting of my research group. It's a time when I catch up on what my postdocs and graduate students are working on, and a time when we all practice describing our work to one another and see if we can help each other out. Often some of the other staff scientists drop by, too.

One day, in a meeting of my research group, I was paying particular attention to the way we were interacting with one another. Have you ever noticed that scientists are perpetually interrupting each other and talking over each other? That's what we were doing in this meeting.

So the next week, during group meeting, I decided to break from the usual here's-this-graph-I-made format and instead to bring up the topic of how we were speaking to each other. I asked everyone: Have you ever noticed that scientists are always interrupting each other? Do you feel like you need to interrupt other people just to be heard?

This question led to an amazing revelation. All of my group—men and women—agreed that there was a critical moment in their careers when they consciously decided to cut people off, interrupt, butt in, and so on. Everyone also seemed to agree that this moment was a step forward. They found more respect from their colleagues once they made the decision to be an interrupter, not an interruptee. This made me feel better; I realized I had gone through the same epiphany. In the company of other scientists, I'm an interrupter too. After I brought up the topic, we all became hyperaware of when we were interrupting each other. It made for a strange conversation, with everyone suddenly feeling cautious.

I did a little bit of research on this topic and I found out that interrupting is not just for astrophysicists. When WomensMedia CEO Nancy Clark interviewed former Secretary of State Madeline Albright and asked her what advice she had for women in the workforce, Albright said, "You need to learn to interrupt. Ask questions when they occur to you and don't wait to ask. Also, you don't need to ask permission to ask a question."[10]

Code Switching and Mimicking

So, clearly, there are times and places for interrupting. But what if you were having tea with your beloved grandmother in a fancy restaurant? Of course you wouldn't interrupt her. You might want to adopt a different manner of speaking altogether to match the company and the setting. In fact, I'm reasonably sure you would do it automatically.

Changing the way you speak depending on whom you're speaking to is called *code switching* by psychologists. Good salespeople not only empathize with their customers—they also mimic them to some degree. Changing their communication patterns helps them project empathy and understanding.

President Barak Obama is an extremely fluent code switcher. An article by Christopher Beam in *Slate* magazine described some of Obama's code switching:

> At fundraisers in New York, he'd put on his professorial lilt. In front of mostly black audiences in South Carolina, he'd warn them against believing rumors that he was a Muslim. "They try to bamboozle you, hoodwink you," he said, in a deliberate homage to Malcolm X. On the Ellen show, he won the week by doing a harmless dance that drove the mostly white audience crazy. After a particularly rough debate in North Carolina, he referenced Jay-Z by brushing dirt off his shoulders.[11]

Like President Obama, scientists and academics are often dancing between one subculture and another.[12] We need to marshal the grumpy lab techs on Friday morning, court the dean's office on Friday afternoon, and then go home to our families on Friday night. That forces us to be code switchers. I want to make clear that this is a good thing—an ingratiating habit that will continue to serve us when we talk to the public or to the press. This technique especially helped me navigate during my trips to Nashville, where I often felt like a fish out of water. Before I showed up in the recording studio, I looked up some dirty jokes to tell. These jokes would never fly at the Harvard Smithsonian Center for Astrophysics. But they put the steel guitar player and the drummer at ease right away. When it was time for me to tell them how to play, they were more ready to listen, and I got a better recording as a result.

Some books I have read carry this idea even further.[13] They argue that mimicking the speech patterns of the people you are addressing is good, but mimicking their body language is even better. When you're in your client's office and she leans forward: you lean forward. If she slouches back and puts her hands on the back of her head: you slouch back and put your hands on the back of your head.

I've tried this approach a few times, and it felt unnatural. But I realized once I started trying it that I'd others do it. Moreover, once I tried self-consciously mimicking the posture and speech of others, I realized that I'd been doing it unconsciously, to some degree, my whole life. As infants, we mimic our mothers' facial expresssions; this mimicking sets a pattern for lifelong social engagement. This is one way in which it might be good for us scientists to become more like infants.

Enthusiasm and Optimism

Steve Jobs, cofounder and former CEO of Apple Computers, has a reputation as a passionate business leader and modern folk hero. In 1999 one of Jobs's friends said, "He is single-minded, almost manic, in his pursuit of excellence."[14] Jobs peppers his presentations with words like "extraordinary," "amazing," and "incredible" (see Table 3-1).[15] When Jobs gave the opening presentation at the computer expo Macworld '08, he began his talk with open arms, a broad grin, and the words "We've got some great stuff for you. There's clearly *something in the air* today." That kind of enthusiasm helps Apple sell 20,000 iPods every day.

Table 3-1 A Few Enthusiastic Words That Steve Jobs Has Used at Macworld Expo, 1998–2008

Stunning
Revolutionary
Incredible
Beautiful
Best
Great
Awesome
Remarkable
Tour de Force
Cool

Of course, enthusiasm is not something they teach in science class. Far from it. Graduate school is all about being tough and skeptical. But as you remember from kindergarten, everybody likes people who are positive and enthusiastic; a smile on your face addresses people's primitive needs for friendship and belonging. So a good salesperson considers optimism to be part of his or her job. To quote Adlai Stevenson, "Pessimism in a diplomat is the equivalent of cowardice in a soldier." Or to quote Anne Kinney, Director of the Solar System Exploration Division at NASA, and one of my mentors, "If you have a method or idea and you believe it works, you have to be optimistic about it. Optimism is the number-one thing."

There's more to it than that; we can actually quantify just how important enthusiasm is. For some reason, negative expressions leave a more lasting impression on our psyche than positive ones. Specifically, negative messages have something like five to seven times as powerful an impact on our minds as positive messages. When a married couple has more than five positive interactions for every negative one, marriage experts say the relationship is healthy. But if the couple starts having fewer than five positive interactions for every negative one, divorce is probably imminent.[16]

Consider the negative words in the table to the right (Table 3-2). Even just looking at them gives me an unpleasant feeling. If you find yourself using them often, people might start associating that kind of unpleasant feeling with you. And it might take five to seven positive interactions to make that bad feeling go away.

If you've ever sat on a review panel or hiring committee, you have probably noticed that if someone says something strongly negative about an applicant, it leaves a lingering stain that can't be erased unless several people override it. For this reason it's important to have at least five to seven members on any decision-making panel. With fewer people on the panel,

Table 3-2 Guaranteed Mood-Killers

No
Bad
Useless
Don't
But
Shouldn't
Can't
Won't
Mediocre
Unintersting
Stupid
Pathetic
Idiots
Sucks
Meh

a single person's bad feelings can swamp the decision-making process, turning it into a black-balling session instead of a thoughtful discussion.

Even when committee organizers take the precaution of assembling a committee with at least five active members, sometimes decisions can get made on the basis of which candidate has the fewest negatives, rather than which candidate is the most outstanding. Nobody wants it to work this way, and shrewd committee chairs may try to combat this tendency. But the powerful effect of negativity is a consequence of human nature that we need to be aware of. If you're writing a proposal or applying for a job, and you expect you might have to impress a committee, your task might be more about eliminating negatives than about dazzling people.

Pride, Modesty, and Humility

Common sales wisdom says that nobody likes a show-off or an egotist; to make friends and to build business relationships, it helps be humble. That might seem like a strange thing to read here after a discussion about projecting confidence and enthusiasm. But a salesperson must always walk a fine line between demeanors: pride, modesty, and humility.

I had a discussion about the importance of being humble with John Mather, who shared the Nobel prize in physics with George Smoot in 2006. He understood this delicate balance. We went back and forth at length about the terminology: humility, pride, modesty, false modesty, and so on. I said that I think you can be proud and humble at the same time. Mather preferred the word "modest" to describe the right attitude for a scientist.

He said that, in science, "the teams you're working on can have ten or even a thousand people. So you have to learn to assume that there's somebody on the team who knows something you don't know. You've got to be able to find those people who can solve a problem you can't solve." That sounds like humility to me—to a Nobel Prize–winning degree.

Should we call it humility or modesty? I'm not sure we exactly resolved the argument over the terminology. In any case, I think I can tell when I've found the right balance. I think it comes from an attitude.

When I have the right attitude, I feel the wealth of what I have and what everyone around me has to offer. I have a happy, comfortable feeling of being surrounded by peers and people I admire. Somehow, when we appreciate ourselves and the people around us, we receive more appreciation in return. That's something John and I were proud to agree on.

CHAPTER FOUR

Building Relationships

 Marketing involves salesmanship; some of it is about painting a picture for people, or satisfying people's short-term needs for positive human interactions. But in the end, the goal of marketing is about something deeper: long-term, honest working relationships. It may seem counterintuitive, but many of today's business experts view all the methods of marketing as tools for building these relationships.

I've come to believe that this view applies just as well to the marketing of science. Despite science's many purely intellectual rewards, human relationships might be the ultimate goal and the ultimate prize for our work. This notion is the single most important marketing concept I want to communicate to you—and the one that I myself had to struggle the hardest to understand.

Many scientists are already wise about some aspects of marketing, and they casually throw around terms like "branding" and "selling yourself." But most comments I hear from scientists reflect an awareness of the craft of marketing as it was practiced thirty years ago. Relationship building has only recently surfaced as a marketing mantra in the business world, so I think most of us are largely unfamiliar with it.

Traditional marketing campaigns from the 1980s and earlier focused more on advertising and pricing: pushing a product out to the masses. There used to be a hard wall between the big companies of the world and their customers, one that dissuaded relationship building—the wall of the television screen. Maybe that's the paradigm most scientists have in mind when they think of marketing.

But when the Internet appeared, a revolution occurred in the business world. Websites like Amazon.com, Tripadvisor.com, and Yelp.com began to feature product reviews from consumers, sometimes right on the page where the product was for sale. Now customers can even read product reviews written by their friends on their cell phones while they are in a store shopping. As you know, these reviews can often make or break a new product.

This situation posed a problem for businesses used to the old way of doing things. How can a mere billboard or television commercial convince you to buy what your friend tells you to avoid? Companies realized that in order to succeed, they could no longer just tell people what to buy; they had to join the worldwide conversation about what's good and what's not.[1] Like a real friend, they had to be present, honest, sharing, and personal. The closer and deeper the relationship they could form with their customers, the better.

One company that understood this new situation right from its inception is TCHO, a chocolate maker founded in San Francisco in 2007. TCHO (pronounced "choh") is led by a group from Silicon Valley that includes Luis Rossetto and Jane Metcalfe, founders of *Wired* magazine. Its business strategy incorporates customers right into the company—exemplifying the new marketing strategies demanded in the Internet era.

At TCHO's start-up, Timothy Childs, the company's Chief Chocolate Officer (formerly at NASA!), realized that there was a flaw in the chocolate-making process. Though consumers of gourmet chocolate prized certain flavor characteristics, the farming and manufacturing processes that generated these desirable traits were understood by chocolate makers only through guesswork. So Childs began installing Macintosh computers with Internet access at Fair Trade farming cooperatives, together with laboratories for monitoring the cacao-growing process. Then, they invited their customers to send feedback to the farmers, to eliminate the guesswork. As the company's website put it, "Since a lot of us have tech backgrounds, we adopted a familiar idea: we encouraged our users to help us make the chocolate they wanted, in much the same way software developers engage beta testers."[2]

The company devised a six-segmented flavor wheel, with labels on it like "fruity," "nutty," and "citrus," and they began to test their chocolate on chefs and customers, asking them to describe the taste using the flavor wheel. They sent the feedback to the farmers by e-mail and asked them to refine their cacao-bean-growing and selection processes to heighten each characteristic on the flavor wheel. There are no nuts in the "nutty" chocolate or orange peel in the "citrus" chocolate, by the way. The flavors are merely characteristics of particular beans, farming methods, and growing regions.

When the founders of TCHO were satisfied that they had iterated and integrated enough input from their first customers, they released their chocolate bars for wholesale as "Nutty 1.0" and "Fruity 1.0" and so on. Their products earned rave reviews from chefs and chocolate connoisseurs. This was almost inevitable, since many of the world's chocolate experts had essentially collaborated in the formulation of the chocolates. That's the power of TCHO's relationship building.

I did not participate in any of the TCHO beta tests, and I have not joined the TCHO "tasters circle" that you can access on its website. But when I open a package and see all the signs of a work in progress where my fellow chocolate eaters are contributing to the flavor, I almost feel like I have a close relationship to the company as well. I picked up a bar of "Fruity 2.0" at Whole Foods the other day; it didn't last long.

With the example of TCHO in mind, I'd like to talk about relationship building as a new pillar of marketing, and also to offer you some relationship-building tools that I think apply to life of a scientist. Though they may draw us out of our comfort zones, I think you will see that they can help us make some wonderful chocolate—and wonderful science, too.

Not a Trick at All

The most important relationship-building tool is simple: you have to be real and authentic. As corporate America is learning, you need to provide products and services that can withstand criticism all across the Internet, or they won't come back. Likewise, you can't use marketing to hide shoddy or dull scientific work. You might say that best marketing practices have come into alignment with scientific ethics. Relationship building is not a trick—it's real.

Besides authenticity, honesty is an important part of relationship building. Maybe all personal relationships start out as superficial or formulaic.

But in a good relationship you eventually find yourself genuinely caring about the other person—and vice versa. You can only achieve this true connection with someone if you keep your promises and you are honest with each other, right from the beginning.

Now, everyone knows scientists who seem to work hard at marketing but mostly just manage to be annoying. Maybe they claim to be first at something when they aren't, or they try to make up a new term for something that already has a name. They aren't honest and real, and you can see right through them.

On the other hand, some scientists just seem to have good relationships with everyone. Maybe they have helped to educate a series of students who all speak admiringly of their wise and gentle advisor. Or maybe they led some large project, but stepped back from the limelight and gave the credit to their hard-working teammates.

I tend to think that in science, the best marketers are the ones you don't think of as good at marketing. The best marketers don't seem to be working at it; they are just plain successful. They may be prize-winners, department chairs, leaders in your field—but whatever positions they have, they seem to deserve their stature. That may be because their work withstands the tests of customer feedback; it's for real.

The Marketing Funnel

Think of your favorite snack food—and all the people who have ever merely heard of that snack food but never actually tried it. There are customers who see it in a store, and consider buying it. There are customers who eat it regularly. There are even customers who buy it all the time and feed it to their friends. Line all those people up in a row, from the many perspective customers to the few addicts and proselytizers, and you have what's called the "marketing funnel" for this snack food.

Relationships seldom start off strong; they take time—and repeated positive interactions—to develop. That gives rise to an important concept in marketing: the need to draw people through your marketing funnel via a series of positive interactions. Companies know that to launch a product, you have to offer something that appeals to customers at *every stage* of the marketing funnel. You need a memorable ad so they hear about the product. You need a nice display in the store to help get them to make

**Never Heard
of You**

Know your work

*A Scientist's
Marketing Funnel*

Collaborator

Advocate

Figure 4-1   A scientist's marketing funnel. Source: Pete Yezukevich.

their first purchase. And you need a loud crunch to keep them chewing them once they've opened the bag.

Now, imagine everyone in the world lined up in a progression—from the multitudes who don't know you exist, to those who have heard of you but are unimpressed, to the smaller group of colleagues who think highly of you, to those few close collaborators and students who are so enthusiastic about you that they would follow you anywhere. That's your marketing funnel as a scientist, and the key to your success. You can think of the rest of the tips in this book as tips for how to draw people through its various different stages.

The year TCHO was founded turned out to be one of the hottest New York summers of recent years. In that oppressive heat, TCHO's John Magazino traveled around New York City, offering tastes of the chocolate to pastry chefs. That was the mouth of TCHO's marketing funnel.

Some chefs were curious. Others were completely uninterested. "Some chefs were so busy they didn't have the time to taste it," Magazino told the *New York Times*.[3] In other words, some chefs never made it past the mouth of the funnel. But others were drawn in by the chocolate's evocative overtones and silky mouthfeel, and they slid deeper into the funnel, eventually becoming proselytizers for the new product. We'll come back to this concept in a moment.

Meeting Someone in Person

Q: How do you tell an extroverted scientist?
A: An extroverted scientist looks down at *your* shoes when he talks
to you!

It's an old joke: scientists are notoriously bad at face-to-face encounters. We might be smart, but we sure are awkward. Always stuck inside our heads! If only there were some kind of formula for what to do when you meet people. Then maybe scientists could just follow the formula and everything would be okay.

Well, there is a kind of formula for what to do when you meet people. It doesn't get you far into the conversation, and it doesn't help you remember the names of anyone's grandchildren or graduate students. But it does carry you through the first few seconds, when you're making a first impression.

So here is one formula for how to greet somebody, adapted from *How to Talk to Anyone: 92 Little Tricks for Big Success in Relationships*, by Leil Lowndes. For dogs, the corresponding formula might involve sniffing and tail wagging. For people, the formula seems to be substantially about eye contact, as the joke above implies.

1) Get yourself in the mood by imagining you are greeting a cherished old friend you haven't seen in years.
2) Turn your whole body toward the person.
3) Pause for one second and thoughtfully study the person's eyes.
4) Now smile a big, warm smile.

I've tried this recipe out, and by golly, it works. Or at least it gets me to the part where I flub up the next part of the conversation. But by then, I've already made a reasonably good impression.

Notice that the formula includes pausing to make eye contact before you smile. That's partly to emphasize the eye contact, and partly for another reason. If you show up already smiling, the person you're meeting might think you're just giving away smiles for free. But if you pause and study the person's face for a second before you smile, it shows that the smile you flash is just for the person you're meeting; it makes that person feel like he has earned it.

Lowndes also says that making initial eye contact is not enough; once you have made eye contact, you must keep refreshing it during your conversation. Of course, steady, persistent eye contact can send the wrong message: it can look like brownnosing or romantic flirtation. But if you deliver a fresh dose of eye contact to your colleague at regular intervals, it will keep your business flirtation going. With the right level of eye contact, you can keep your scientific relationship hovering in a sweet spot somewhere in the range of "friend" and "business associate." As Victoria McGovern suggested in an article on the Science Careers website, "Don't glare like a vulture, just make eye contact—'check in' often to see if his face registers understanding, engagement, or a strong desire to ask a question."[4]

Some scientists seem to believe that to be taken seriously, they must be perpetually serious. Then they forget to smile, which leaves them and the people they meet feeling dour. To me, the formula above combines the best of both worlds: a serious gaze and a wide smile. The combination shows that you are intense, thoughtful, and hard to please, but also a potential friend—maybe someone it might be fun to make chocolate with.

The Courtship Analogy

Imagine one day you received a letter in the mail from some company you've never heard of, and the letter asked you to buy their product, and to tell all your friends about how good that product is. Of course, you would throw it in the recycling bin. That's not good marketing.

Marketing guru Seth Godin likes to make an analogy between moving a customer down the marketing funnel and the gradual process of courtship between two people.[5] The idea is that when you are building a relationship, you have to take it slowly, one step at a time. In the example above, the company that sent you the junk mail was moving too fast—Godin might say that this blunder was like asking you for your hand in marriage before even asking your name.

Let's say there's someone you want to develop a relationship with, as though you were Magazino from TCHO chocolate courting a New York chef. The first step is making contact. You might start by sending an e-mail or having someone introduce you. Then, ideally, you'll try to set up a meeting or at least a phone call—a kind of first date. "Hey, do you mind if I drop by and give you a taste of this new chocolate I made?" Even just getting to this stage requires some patience, luck, and empathy.

The next big step in the relationship might be where one of you does the other some kind of favor. Both the act of giving and the act of receiving involve some kind of risk; they show that there is some level of mutual trust. You might give the chef a sample of your tasty chocolate.

At some point, if you are lucky and your chocolate is good, you might win a convert, someone who will take a risk based on your product. The chef might try serving the chocolate in her restaurant. That's a triumph. But there's more work to be done. From this point, the relationship will need nurturing to maintain and mature it: some kind of contact now and then. Sometimes it just takes time to earn someone's trust. That's why TCHO has a blog and a "taster's circle" club. And that's part of why they keep updating their products, from 1.0 to 2.0, showing that they are still listening to their customers.

What confused me about trying to apply the courtship analogy to science is that sometimes, while scientists are getting to know one another, they may be arguing. We scientists take pleasure in hashing over the fine points, trying to cover all the bases, getting ever closer to the truth of the matter. For some of us, this process is more satisfying if it involves raised voices or a long series of ever-more-picky e-mails.

To make things more confusing, some scientists clearly enjoy locking horns in an argument, but others do not. Some scientists can even take offense if you begin to question their work, and they'll pull away.

But I still think that the courtship analogy still holds. It's just a matter of being open to a range of possible modes of interaction—experimenting with them bit by bit as you flirt. Let's talk about a few situations where I think this analogy works for scientists.

Taking It Slow

It takes me months to write a song: gradually composing it, then refining and editing it. I write maybe fifteen songs in the course of a year and then pick three or four of these that I think country artists might want to record. Then, when I make a trip to Nashville, I usually spend a few days in the studio, recording demos of these songs. We might work into the night bringing the songs to life, overdubbing tracks, adjusting the mix. Finally, the sounds I've been hearing in my head all year turn into something real that other people can listen to. I can hardly sleep because I'm so wound up.

It's tempting, at that point, to spend the next few days wandering around Nashville shoving CDs into people's faces and pockets. Hi, nice to meet you! Want to hear my music? But one of the first rules of Nashville etiquette is that this practice is strictly uncool. That's because anyone in the music business is constantly inundated with such offers. If you start pushing your CD on people you've just met, you instantly mark yourself as a neophyte.

Now, sometimes when I am at a scientific meeting I feel the same way as I do on my whirlwind trips to Nashville. I've been working all year on this paper: I want you to read it, now. Or sometimes I'm on the other side of the desk: an eager postdoc shoves a thick stack of her latest papers into my face. That's an uncomfortable moment. And despite my initial intentions, these papers usually end up crushed in the bottom of my bag, and then in the recycling bin, unread.

The flirtation analogy helps us find a better approach. A flirtation starts slowly, gingerly. Each person checks carefully with the other, asking for permission to proceed. In a romantic flirtation, you might ask, "Would it be too soon to ask you have to have dinner with me?" In a scientific flirtation, you might ask, "Would it be okay to send you a copy of my paper?"

In Nashville, you are expected to proceed even more gingerly, and wait till someone asks you about your music. It's assumed that if you are in Nashville, you must have some music to pitch—everybody in Nashville is a musician. Once people determine you aren't going to shove a CD in their faces, they are often polite enough to ask you about your latest recording.

Schmoozing

We can take the flirtation analogy further. If you were on a first date, you would probably try to get your date to talk about himself or herself. You might ask, "So where did you grow up?" "What did you think about that movie?" That's because when you ask someone for advice or for an opinion, you gain twice. First, you might learn something valuable about the topic at hand or the person you're speaking to. Second, you show that person respect by indicating that you value his or her opinion.

Scientists always seem to enjoy describing their work, offering their criticisms, pontificating, and so on. But not every scientist has learned the double sweetness of lending a receptive ear and having a conversation about what someone else has to say. This is another relationship-building technique I think we can easily benefit from practicing more often.

This simple tool—listening and responding—works both offline and online. Scientists love to post things on the Web for all the world to see. But it seems relatively rare that a scientist actually responds to someone else's posts, comments, etc. Without showing interest in other people and engaging in conversations about what they have to say, you're just an old-fashioned company with a billboard and a prayer—and modern companies like TCHO are poised to steal your customers.

Topics of Conversation

A related piece of business advice is that it helps to chat about topics other than business. The idea is that you might find you have common ground besides your business interests—common ground on which to build a stronger relationship. And when you chat with someone about nonbusiness topics, you show that you value her as a whole person, not just a means to an end.

For example, you might want to ask someone, "What else do you really love to do, besides publishing country songs?" Or sometimes you can guess: "Is that a blowfish-diving suit sticking out of your backpack?" Any hobby you might have in common is something you can build relationships around.

But here's one place where the business world and the science world seem to part. In the business world, a relationship starts with small talk, and then shifts to business. In science, the pattern seems to be reversed. At a scientific meeting, it's perfectly permissible just to talk science. Or you start by talking about science. Then you can introduce other topics into the conversation once you get to know someone.

Meeting Many People at Once: Throwing a Party

The Simons Lab (I changed the name, and some other details) had a striking, modern-looking facility in a suburban area—and a steady need for funds. Tantalizingly close to the lab was Oak Heights, a wealthy neighborhood of big houses with lush, gated gardens. Surely the millionaires who lived in those houses had a bit of money to spare that could make a big difference to the lab's budget. But these wealthy people seemed to hide behind the tall garden gates, close yet out of reach.

The lab's public affairs officer—I'll call her Karen Bridges—had a clever idea about how to pry open those gates. She set aside an October evening, called a high-end caterer, and began to prepare for a gala reception at the lab. She purchased a mailing list of all the people who lived in the nearby wealthy zip code, and printed embossed invitations on thick paper. She placed the invitations in large, gilded envelopes, and dropped them into every single mailbox in the neighborhood. The invitations announced a "Special Gala Evening, Just for Our Neighbors in Oak Heights." She rented lighting and hired a harpist. Fifteen thousand dollars later, she had the whole event planned.

At first, there was no response to the RSVP on the invitation, and Bridges began to get nervous. But most of the wealthy citizens regularly drove past the lab's curiously asymmetric building, and some of them wondered what it might contain. One retiree had nothing better to do that night, so he responded affirmatively to the invitation—mostly to see what was inside the remarkable building. A housewife's family was out of town; she decided to drop by and to bring along a friend. Little by little, the event gained in popularity, till suddenly it became the hot ticket for the weekend. Thursday before the event, Karen counted the RSVPs and realized she had a full house.

On the evening of the gala, the air was crisp. Tiny lanterns lit the trees, and servers in tuxedos waltzed through the crowd with pumpkin-scented canapés and champagne. Amid all that luxury, Bridges could almost feel the money draining from her account—money that could otherwise have funded a graduate student for most of a year.

The director of the laboratory gave a stirring presentation about the laboratory's work, boasting about all the dedicated people who worked there and all the good their research would do for the world. The crowd, enraptured by the spell of the evening, listened intently. Near the end of his talk, the director added a few slides breaking down where their funding came from, and describing the wonderful work the lab could do with just a few million dollars more. He asked the crowd for their help, offering to place the names of any donors on a large permanent-looking plaque in the lab's foyer.

I don't have to tell you that this scheme paid off handsomely. The lab raised tens of millions of dollars, far more than the cost of the party itself. It generated a new mailing list of potential donors that has now served the lab for almost a decade. In fact, since that evening, the donors themselves have begun throwing parties to raise funds for the lab.

The moral of the story is that throwing a party, or going to a party, is a great way to meet people and build relationships. It works for seven-year-olds, millionaires, and scientists. You might literally have a party at your lab or at your home, and invite strangers, or colleagues. Or if your goal is to build relationships with scientists, you could throw another kind of party: you could host a conference. Of course, the scientific conference is a deeply entrenched academic institution. But to a marketer, a scientific conference is essentially a party: an ideal way to attract new clients and keep up with old ones.

If you were throwing a party, you would go out of your way to make your guests feel welcome and appreciated. You would send out polite invitations, and you would graciously tolerate guests who were late to respond. You would serve delicious, colorful food. You might have party games or serve alcohol to help loosen people up.

And a scientific meeting or conference works the same way. The idea is to make your guests feel special, like they have come to an exciting event. We will talk about this analogy more in chapter 10.

The New Person

When I first showed up in Nashville, I expected there to be some kind of culture shock. For example, I grew up in the Northeastern United States (New York, Boston, etc.), whereas Nashville is clearly in the South. And even though I braced myself to be misunderstood at every turn, it was worse than I anticipated.

I eventually learned that this is because Nashville is a melting pot. People come to Nashville from all around to pursue music. From a distance, Nashville had looked to me like "the South." But Georgia, East Texas, West Texas, Kansas, Louisiana, and West Virginia all have different cultures, mannerisms, dialects, and customs. All of those are represented in Nashville's music community. There's also country music in Wyoming and Oregon and Canada (and even, say, San Francisco and New York City), and pilgrims from all of those lands and many others make their way to Nashville. I even met one group of writers who had moved there from England to write American country music.[6]

Fortunately, there were people who reached out to me when I was new and clueless, and they taught me some of the ropes. Some writers, musicians, and publishers were too busy to spend time with me, or rejected me when I didn't fit in. But a few people kept an eye out for me, the new guy. These are people I've never forgotten, and I still refer business to them.

So now it's my turn: I try to keep an eye out for new people—the new faculty member, the new graduate students, the new staff member at the funding agency, even the freshman senator. They are going through a crisis, so you have an opportunity to help them. It's not obvious if or when your help will be repaid. But the fundamental theorem of marketing tells us that helping other people is the only way to get what you desire. And it feels good, too.

I remember one day in Nashville I was dropping off a CD and I suddenly found myself in a basement recording studio full of tattooed men sitting on the desks, spitting tobacco—not my usual crowd. I felt my heart climbing up my sternum. When you first show up at a scientific conference it can feel just about the same. If you're uncertain what do to at a meeting, you can always find someone who feels more uncomfortable than you to talk to: the person who just walked in. Likewise, in any Internet group, there's always someone hoping you'll reach out to him or her: the person who just joined.

The Benjamin Franklin Effect

In 1966, a team of psychologists phoned housewives in California and asked them for a small favor: Could they answer a few questions about the household products they used? Some agreed to answer; others did not.

A few days later, the psychologists called those same housewives again plus a second group of housewives whom they had not talked to before. This time, they asked for a much larger favor. Could they send a team of five or six men to visit the houses and go through the cupboards and storage places, enumerating household products? The psychologists found that the women were more than twice as likely to agree to the bigger request if they had already agreed to the first request.[7]

When someone does you a favor, you might think at first that you have become indebted to that person—and politeness and social conditioning dictate that this is true. But another strange thing happens in this circumstance that's far from obvious: when someone does you a favor, that person actually likes you more, afterward, and is more willing to do you further favors. This phenomenon is sometimes called the Benjamin Franklin effect, because Franklin described it in his autobiography: "He that has once done you a kindness will be more ready to do you another than he whom you yourself have obliged."[8]

Here's how it works. If we do a favor for a stranger, we justify this action to ourselves. We tell ourselves a little story: I am a thoughtful person, and my efforts were well spent because I *like* that person I just helped. The story we tell ourselves is the story we believe. And so we help the story become true.

I mention this idea, of course, because as the business books will tell you, asking for a favor is one way to help move someone down your marketing funnel—though it may be a paradoxical, counterintuitive way. You can even encourage someone to do you a big favor by first asking for a small favor, as the experiment with the housewives shows. The books on business relationships are full of fun tidbits like these.

The Benjamin Franklin effect has another, more sinister side. If someone does something to harm you, he becomes more likely to do so again. This effect occurs in the same manner we just discussed: people tell themselves a story that justifies their actions. If they harm you, they tell themselves that they harmed you because you are an unworthy person. Then this dislike becomes their reality.

That's something to avoid, of course. So the advice books say that if you sense that someone is going to be forced by convention or agreement into doing something to harm you, it's in your interest to stop the process at once to prevent breeding dislike. For example, it is better to say farewell rather than be asked to leave, better to keep your presentation short rather than be asked to end it, and better to resign than to be fired.

Rejection

In the songwriting business, there is a saying that in order to do a song justice, you need to pitch it at least a hundred times. It's just as bad if you are with a band trying to get signed to a record label. When Decca Records rejected the Beatles, Dick Rowe at Decca told their manager, Brian Epstein, that "guitar groups are on the way out" and "the Beatles have no future in show business." Instead of the Beatles, Decca chose to sign a band called the Tremeloes, who were local and would require a smaller travel budget.

Science seems relatively gentle by comparison to the music business. The first time many scientists face serious professional rejection is when they fail to get a fellowship or a faculty position. It comes as a shock. You've racked up an long string of academic sucesses: acceptances to schools and special programs, scholarships and prizes. You were assured that there would always be a need for someone with your talent, intellect, and interests. You are safely embedded in your ivory tower thinking deep thoughts. Suddenly, doors are slammed in your face; the Tremeloes made the shortlist instead of you.

When I started applying for jobs in science and started getting rejected, it dredged up memories of being rejected by kids on the playground, rejected by women, rejected by family. The pain just kept going on and on, deeper and deeper, churning into a vortex of self-doubt and frustration. I even wondered if part of why I had decided to try science—and then the even more frustrating business of songwriting—was that I somehow craved the rejection.

One of the major lessons of marketing is that if you don't succeed, it doesn't necessarily mean you are bad or mediocre—it just means that you haven't met anyone's particular needs yet, or convinced someone that you are the best person to meet their needs. There are many, many consumers in the marketplace, and one of them may need your product—someone whom you haven't seen recently or perhaps haven't even met yet. Or you might be able to tailor your product to meet someone's needs once you learn more about what those needs are. Nobody doubts the genius of the Beatles, but even geniuses meet closed doors.

Well, maybe that notion isn't so comforting. But here's another take on the situation, one that I live by. If you want to go anywhere interesting in life you *have* to get rejected. It's a function of how ambitious you are. If you are working in a situation where everything you want arrives on a silver platter, you've set your sights too low. If you aim higher, you're sometimes going to miss your target. You might even say that successful people succeed *because* they were able to take risks and face rejection.

That viewpoint works for me. I'm an achievement addict, and I bet you are, too. If I quit science, and went into some other line of work, I know I would find myself chasing some other goal just out of reach. And there I would be again, facing rejection after rejection. I suppose I could back down and settle for a simpler life, but mostly that's just not me. So I've come to welcome the rejections, and even the occasional blues they bring, with a sort of perverse pride. (By the way, the book you are reading now was rejected by sixteen different publishers before the folks at Island Press recognized its potential and inner beauty.)

Now, many people—scientists included—are bad at handing out rejection. Instead of saying, "no thanks," they will just stop returning your calls. Or maybe you'll get a formulaic rejection letter two months after everybody knows you've been turned down. This common lapse in social grace is the basis of a recent book, turned into a movie, called *He's Just Not That Into You*.[9]

But here's another lesson I learned from the music business. If a customer is not returning your e-mails and phone calls—and after a while you start to get the message that this means no—you do not need to give up. I've had people pass on songs I pitched them for their first album, then cut the very same song that they passed on for their second album when I pitched it again. So if someone tells you no, you can just put him down for a "no, thank you" and give him some space. Then if you are still interested in this customer, you can come back later, be polite and considerate, and roll the dice again.

So if you haven't yet made a deal, then you may have nothing to lose by continuing to pitch your ideas to someone, even if they have rejected them before. "No" doesn't always mean "You're a bad person and I hate you forever." Sometimes, it just means "No thanks, not this time."

The trick is to be sensitive to the range of "no's" by attending to other nonverbal cues as best you can. A "no" with a handwritten note might be an invitation to continue the conversation. A "no" from an underling might mean the message hasn't gotten through. In any case, regardless what answer you hear at first, as Winston Churchill said, "Never, never, never give up."

Congratulations!

Rejection may be hard. But success can be hard, too, in a curiously similar way. Both rejection and success can have a way of temporarily reducing me to a kind of childlike state. Do you know what I mean?

For example, when I walk through the door after making a great discovery or giving a great talk, I just want to hear someone say "WOW"! Maybe inside every one of us there is a two-year-old undergoing potty training, craving a hit of positive feedback for a job well done. The two-year-old in me didn't really go away when it got a PhD.

From the standpoint of marketing, this need for instant approval is just another human need, and one that's relatively easy to meet, as follows: if someone does something that's really super and impressive, give him or her a big "Wow!" or "Nice Job!" Almost any word will do as long as it has an exclamation point on it. It's also got to come right away after the announcement—instantly, if possible. I don't think you have to worry if the discovery or the announcement turns out to be wrong, or ultimately unimportant. Everyone will understand that you're just reacting to the emotion of the moment.

Scientists are generally bad at giving each other praise, fearing perhaps that they might be seen as having low standards, or sucking up, or overcompensating for something. And our friends and families often don't understand what we're doing well enough to praise us in proper proportion to our achievements. So we're often hurting for positive feedback. I've found that giving my colleagues a little bit of well-timed, vigorous attention can go a long way toward healing the hurt they feel—as all scientists do—for this lack of understanding. It might seem like a frailty, but I would want you to give me this attention as well.

Scientists are also lousy at accepting compliments. I know I am. If I've really just knocked the ball out of the park, I might be feeling excited enough to high-five someone. But as soon as the high wears off from whatever it is I might have accomplished, I get self-conscious. Sometimes I might pause awkwardly or even dispute the compliment. Someone might say, "Hey, that was a great paper you wrote!" And then I might say, "No, it wasn't, really. There were some mistakes in it—we just didn't get them fixed in time." Then the person who complimented me feels awkward. And I feel annoyed. So then we stand there, the two of us, feeling miserable together.

But it pays to receive compliments, and pays even more to receive them graciously. When others see you receiving praise, they develop a higher opinion of you. To take full advantage of this effect, we have to make it enjoyable for people to compliment us by receiving compliments in the spirit in which they were given—because someone who enjoys praising us once becomes more likely do it again. The positive feedback can turn into a spiral of increasing goodwill.

For my own benefit, I checked out a few different books about relationships and looked up the right thing to say when someone congratulates you.[10] The right answer is simply something like "Thanks, that's very kind of you" or "It's so wonderful of you to say so." You can also take the opportunity to give credit to the other people who were involved. And then, it's important not to forget that you're in the middle of a conversation—it's probably time to turn the attention to the other person. "And how are you?"

Staying In Touch

Another vital part of relationship building is is staying in touch: finding reasons to stay in contact over the long term with people you haven't seen in a while. It's amazing how time can make a relationship feel deeper. That's one reason why the folks who sold me my car keep track of my birthday; it's an excuse to stay connected. I don't especially enjoy hearing from Toyota on this occasion. But, of course, there is a whole range of real friends I do love hearing from on my birthday.

A more professional "excuse" to stay in touch with someone is the social obligation to send congratulations when you hear about someone's success, as we were just discussing. It's good to send colleagues an e-mail when you hear they got a degree, published a paper, won a prize. It's good thing to send congratulations when a policymaker has gotten some new legislation passed or a reporter has written a prominent cover story. We'll talk more about staying in touch in the chapters on conferences and Internet marketing.

A Philosophical Comment

Scientists and other academics have created for themselves a community that we sometimes view as a kind of meritocratic utopia. In science, family background and personal wealth don't seem to matter as much as they do in the rest of society. On the surface, interpersonal relationships seem to count less in science, too. It might seem superficially like one's status in science might be based purely on talent and expertise, and that relationship building ought not to matter.

But the notion of a scientific meritocracy falls apart at the seams where subfield meets subfield. It crops up in the department meeting when the physical oceanographers insist that the next faculty hire must be a physical oceanographer. It emerges when the conference organizers haven't read your paper yet because some other topic is hot this week. No degree of intelligence or expertise will convince some scientist who has never really heard of your work to kowtow to your expertise in these situations.

And I think it's even messier than that. To me, scientific thought resembles some kind of fractal, with boundaries everywhere on all different size scales—much like a sponge.[11] There are walls between universities, walls between research groups, walls between one scientist and another. In a way, every research group is a kind of sub-subfield, where the same rules apply. You might even say that every good scientific paper, together with its authors and readers, forms its own sub-sub-subfield.

Each wall between scientific subcultures can only be penetrated by a real, organic human relationship. You need to know someone who can translate the jargon and decode the hidden biases. And you can't stay completely sequestered in your own research group forever; you'll always be trying to expand through the boundaries around you. Forming and managing human relationships, it appears, is an inseparable part of a career in science.

Networking vs. Relationship Building

Not all scientists find the need to work on relationship building to be new and strange. One graduate student, Erika Nesvold, told me that her generation takes it for granted. She said, "They keep advising us to network, network, network. After a while it becomes redundant. Because we are networked. We're the generation that invented Facebook."

Networking is now quite the buzzword in career workshops for scientists. At one such workshop that I attended, every student at the meeting was handed a notebook and encouraged to collect signatures in it from other conference attendees. If you got twenty signatures or more, you could win a prize. This game was meant to teach the students about networking—or perhaps just help people get to know each other.

But it seems like there is a gap between the notion of networking as taught in career workshops and the notion of relationship building I've tried to describe in this chapter. Meeting people is a good start. But it is not the same as building a list of potential clients, asking permission to interact with them, flirting with them intellectually, and gradually nurturing the kind of long-term collaboration and mutual admiration that brings both of you joy and professional fulfillment.

Maybe this idea of relationship building is hard to teach because it's ultimately a long, slow process. You can't demonstrate it in the course of a one-afternoon workshop any more than you could have a first date, get engaged, and then get married in the span of a week. But maybe there are some ideas about maintaining and nurturing relationships that we can gradually teach and demonstrate to students. One scientist I talked to about this problem, physics professor Markos Georganopoulos, suggested, "Instead of just talking about networking, scientists can lead by example; they can have social events in the department. Students see faculty that never leave their offices, and they copy them. But if we can hold regular department events with some beer, or get together for lunch, we can teach students to be better at developing relationships with other scientists."

Having regular departmental social events does seem like a good way for senior scientists to model the relationship-building behaviors that younger scientists need to learn. Maybe not all scientists want beer and chit chat when they get together. But the point is that it's crucial for young scientists to learn about *getting together,* whether it's for beer and chit chat, or just to do work. As Nobel Laureate John Mather told me, "Science is a very, very social enterprise in a way that the general public does not expect." Young scientists and science students may also lack this expectation.

Pushing vs. Pulling

I used to think that marketing meant pushing: pushing my ideas outward from my desk to the eyes and ears of the public, pushing through walls of apathy or disbelief. But in the process of writing this book, I learned a different point of view. Marketing is more about pulling people close to you: attracting and alluring people, bringing them from outside your social circle to inside it, ushering them along from distant stranger to close collaborator. This change in perspective helps me relax somewhat: it's easier and happier not to have to push all the time.

In this chapter, we looked at a variety of ways to pull people in: meeting people and moving them along from acquaintence to friend to long-term partner and proselytiser. But how can we pull in people with whom we have no personal connection? How can we influence a group of strangers that's too large to meet in person? And how can we keep pulling on people even when we aren't interacting with them? The next chapter will answer some of these questions.

CHAPTER FIVE

Branding

 They say there are eight hundred songwriters in Nashville who have had number-one hits. That makes sense if you do the math. Let's say there are roughly ten number-one hits per chart per year. Let's consider just one chart (a Country chart), since we're just talking about Nashville (not New York or Los Angeles, for example). Each hit typically has one to three writers on it; call it two. And writers live maybe half a human lifetime after they have a hit—say 40 years. Ignore the small number of people who have written more than one number-one hit and that puts the number of Nashville songwriters with number-one hit songs at 40 X 10 X 2 = 800.

Eight hundred songwriters who have had number-one hits. That means if Kenny Chesney, the country megastar, wants to have a team of twenty number-one-hit songwriters in his living room, he can set that up in an instant. And he can fire them all the next day without fear of hurting his options. Then there's an even larger army of other writers with hit songs that only peaked at positions 2–40 whom he could have waiting in the wings. How could a new writer hope to stand out among such a talented crowd?

At some point, I realized that to distinguish myself from the other writers in Nashville; I needed to brand myself. The word "brand" originally meant a symbol burned onto a cow's hide to identify the ranch it came from—to distinguish the cow from many thousands of others. Companies use the modern concept of branding every day to market themselves. When you want ketchup, you reach for Heinz. When you are about to sneeze, you reach for a Kleenex.

It was almost obvious what direction I should look to find my unique calling card. Of those hundreds and hundreds of songwriters, how many could possibly be astronomers? I knew many astronomers/astrophysicists (those two designations are approximately interchangeable, by the way) who are good musicians, and even several who write songs; people had introduced us over the years. But I was reasonably sure that I was the only practicing astrophysicist who wrote in the style of contemporary country.

So I set out to brand myself as the songwriting astronomer. I chose a logo for myself with a star in it, and bought address labels and a color printer, and began labeling my CD packages with the logo and a deep-blue field to remind people of the night sky. I wrote myself a bio highlighting the connection between my songs and my science. It told how I would bring my guitar to the observatory with me and write songs on cloudy nights. I put it up on my website and included it in my promotional packages. Each time I went to Nashville I brought with me a bag of astronomy souvenirs, handing them out to people I worked with.

Because my brand was based on the truth, I found out that it had a kind of coherence that I didn't initially expect. My songs, it turns out, were somewhat more harmonically complex than typical country-radio fare. At first, other songwriters criticized me for breaking some of the rules, rules about not adding too many color notes to my chords, for example. But my brand turned these infractions into a feature; the studio musicians, when they were struggling to make sense of the harmonies in the bridge, cracked jokes about how I was giving them more of those "astrophysicist chords." Some artists and publishers might have been turned off by my harmonic choices. But others complimented me on being progressive—bringing them closer to the "future of country music."

In science, brands might be more subtle than they are in the entertainment industry. But recently, when I was organizing a scientific conference, I became quite sure that I was seeing a bevy of scientific brands in action. When you're on a committee discussing who the invited speakers should be, you notice how everyone almost instinctively names the same two or three people. These scientists may or may not have made the most important contributions to the field during the last six months. Sometimes these scientists even have somewhat negative brands, but they are needed at the meeting anyway to stir up controversy. What's important is that they are familiar; they are on everyone's mind because they are the go-to people for some topic or angle. That is branding at work.

This brings us to the modern meaning of the word *brand* in the context of marketing. A brand is the set of all expectations consumers have about a company or product. Or, as Marty Neumeir said in his book *The Brand Gap*, "A brand is a person's gut feeling about a product, service, or company."[1]

A brand is a product's promise, its reputation. If all products were pretty much the same, like the cattle on the open range of the American West in the nineteenth century, then the symbols stamped on them wouldn't mean much about their quality. But we suspect that all products in today's vast intellectual marketplace might be different, and we fear the unknown. So we tend to choose the ones we recognize—expecting things will turn out well.

Songwriters carry around two kinds of brands. They themselves *are* brands as songwriters; they have reputations for writing good (or bad) songs. In addition, each song they write is a brand; a hit song, with its promise of emotional impact, can sell many concert tickets and other products.

Likewise, our first brands, as scientists, are our own reputations. I interviewed David Pinsky, scientific director of the University of Michigan Cardiovascular Center, and asked him about branding. Dr. Pinsky has a reputation as one of the country's foremost experts on the relationship between blood flow and diseases of the heart and brain. He told me, with appropriate pride, "My name is my brand: This is work from the Pinsky lab. The branding is my last name." Sometimes this brand, the brand of your own reputation, is called your "personal brand."[2]

And, just as songwriters do, scientists also create other brands. Astronomer Lyman Spitzer is sometimes called the "Father of the Hubble Space Telescope." Computer scientist Radia Perlman has been called the "Mother of the Internet." Arthur Gordon Webster has been called the "Father of the American Physical Society." The Hubble Space Telescope, the Internet, and the American Physical Society all have their own reputations, and are all wonderful brands, promising to deliver good things to those who buy into them.[3]

Now, everyone and everything has a reputation, like it or not. So every person and every product has a brand. The trick is to develop *effective* brands that people remember and admire.

Developing an Effective Brand

Part of developing an effective brand is spreading the word about it. But there's much more to it than that. To create a brand, you've got to create emotions, feelings. As Walter Landor, founder of the legendary brand consultancy Landor Associates, famously said, "Products are made in the factory, but brands are created in the mind."

For example, do you feel just a little bit uncomfortable when you read the words *string theory*? String theory, as I'm sure you know, is the name given to the ongoing discussion of a certain class of mathematical frameworks that attempt to encompass both quantum mechanics and general relativity. It is intensely sophisticated mathematically, but has yet to yield a single prediction that is accessible to experiment. I find that because of string theory's mixed reputation—some say that because it is untestable, it is only marginally scientific[4]—I have a slight visceral reaction to seeing the word or hearing it said. That visceral reaction helps make it an effective scientific brand.

String theory is a research question. The uncomfortable feeling it generates is part of the brand's power; many scientists stay employed trying to resolve that feeling. You might say this uncomfortable feeling is the scientific equivalent of "Ring Around The Collar," the peculiar stain that was first named in order to sell Whisk! laundry detergent.

The *Hubble Space Telescope*, on the other hand, elicits a warm fuzzy feeling. After all, it was the hero of a high-tech drama—a victim of an optical alignment problem and then brought back to life by astronauts. The word "Hubble" feels comforting, like "cuddle" or "bubble bath." And nobody can forget the long string of astounding images of nebulae, galaxies, planets, and so on that Hubble has brought us over the years. The telescope's strong brand helped save Hubble from being axed by budgeters more than once during its lifetime.

Shortly, we'll talk about the standard tools for developing the image of a brand: logos, brand names, and so on. But now I want to take a look at the philosophy of branding—for creating a reputation in the mind. Al

Ries and Jack Trout worked in the advertising department of General Electric and then left to build their own marketing firms. Together they wrote a string of classic, best-selling books on marketing and branding. We discussed one of them earlier, *Positioning: The Battle for Your Mind*. Let's look at another one of Ries and Trout's books, *The 22 Immutable Laws of Marketing*, which summarizes many of the fundamental principles of branding.[5] Not all of the 22 "laws" really apply to science—but several do seem germane and even central to our profession. Let me run a few of my favorites by you and you can see if you agree.

Ries and Trout Law #1: Get There First

Ries and Trout begin by pointing out that the best way to launch an effective brand is to be the first at something. Being first seems to be the password needed to write into certain sectors of the human memory system. For example, everyone remembers Charles Lindbergh. But who was the second person to fly solo across the Atlantic?[6] Of course, scientists are always striving to be the first at something. So I know you don't need to be reminded about this law. But you might find it interesting to read examples of how it has worked in business.

Kleenex, everyone's favorite example of a successful brand, was invented and first marketed as a means to remove cold cream, a kind of makeup remover. The original Kleenex advertisements featured Hollywood stars who used cold cream to remove their theatrical makeup and then used Kleenex tissues to remove the cold cream.

Then, in 1926, the Kimberly-Clark corporation, which manufactured Kleenex, started getting letters from people who were using Kleenex tissues as disposable handkerchiefs for blowing their noses. So Kleenex changed its marketing strategy and re-branded itself. Kleenex became the first disposable handkerchief on the market—and an early example of a product invented by its customers, not unlike TCHO chocolates.

Ries and Trout Law #2: If You Can't Be First in a Category, Set Up a New Category You Can Be First In

As noted above, we scientists are always trying to be the first to do something. But of course, it doesn't always work out that way. Nature doesn't always cooperate. Or it might take ten years for your project to succeed, and by the time it does, other people's projects might have overtaken yours.

So let's say you didn't discover the first extrasolar planet. Instead you discovered the 352nd extrasolar planet, a nifty little critter named HAT-P-11b. As a practitioner in the field I certainly appreciated hearing about this discovery. A team led by Gáspár Bakos discovered HAT-P-11b using a relatively new technique called the transit technique.[7] The planet is about the size of Neptune, and one of the first planets in this size range to be discovered; most known exoplanets are much larger. But Ries and Trout remind us that you won't get widespread applause for announcing that you discovered the 352nd extrasolar planet, despite how hard you worked on it and how interesting the planet may be to specialists like me.

Ries and Trout would say that to remedy this problem you need to find or invent a new category you can be first in. For example, maybe your planet is the first planet discovered by the transit technique. Maybe it's the first exoplanet to be discovered that's just about the same size as Neptune. If you can make either of these claims, then that's what you want to highlight in the press release: the word "first." And don't mention the number 352.

Ries and Trout Law #5: The Most Powerful Concept in Marketing Is Owning a Word in the Prospect's Mind

I'm skipping laws three and four, as well as several others, because they don't seem to be relevant. But of all Ries and Trout's various "laws," number five is the one that my postdocs and students are sick of hearing me tell them over and over: to launch a brand, it really helps to own a word. Obviously, part of defining a brand is coming up with a word that names it. And Ries and Trout give some examples from the business world of companies that have succeeded in attaching themselves to existing English words. What shipping company comes to mind when you hear the word "overnight"? (Federal Express.) Which ketchup comes to mind when you hear the word "slow"? (Heinz.)

The crazy thing is that scientists can achieve both goals at the same time—making up a brand name and owning a word. That's because our society gives scientists special license when it comes to making up words; when we make up a brand name, it can quickly become an actual word in the English language. That's a phenomenon and a responsibility worth pestering your research team about.

Let's return to HAT-P-11b, the 352nd extrasolar planet to be discovered. HAT-P-11b wasn't the first exoplanet discovered by the transit technique. It wasn't the first extrasolar planet discovered that's about the same size as Neptune. It wasn't even the first extrasolar planet discovered by the transit technique that was the same size as Neptune. It was close to achieving all those distinctions, but not quite. It was the 352nd extrasolar planet.

But the discovery of this planet garnered widespread news coverage; the story was picked up by National Geographic news, *Astronomy* magazine, and many other outlets. That's because the press release about HAT-P-11b helped coin a new term; HAT-P-11b was the first planet referred to in a major press release as a "super-Neptune." And that name caught on. If you Google the term "super-Neptune," HAT-P-11b is still the planet you'll find mentioned most prominently in the results, even though by now there are many other known exoplanet planets that could be described this way.

Scientists make up words like this all the time. Deborah Leckband, Reid T. Milner Professor of Chemical and Biomolecular Engineering at the University of Illinois at Urbana-Champaign, gave me another example. "People often decorate micelles, microscopic oil droplets, with strands of polymers. [One chemist] started calling these micelles modified with the long polymer strands 'hairy micelles.' And now that's what people call them!" I think you'll find, if you haven't already, that coining new words can be an extremely useful way to encapsulate a new idea and help others realize that it *is* a new idea.

I've certainly felt sometimes that coining a word has some stigma associated with it. For example, some people refer to newly invented words as "buzzwords" when they want to be derogatory. Maybe you're worried, as well, that coining a word might make you seem self-promoting or conceited. Maybe we should indeed be extra cautious here, since if you as a scientist make up a word, it might become an actual part of the language, not just an ordinary brand name.

But ultimately, every scientific term was made up by one scientist or another. In a way, you might say that the history of new ideas in science is but a string of word coinages:

cell (Robert Hooke)

nucleotide (Phoebus Levene)

calorie (Nicolas Clément)

quantum mechanics (Max Born)

osmosis (Thomas Graham)

electron (George Johnstone Stoney)

neuron (Heinrich Wilhelm Gottfried von Waldeyer-Hartz)

species (Ernst Mayr)

seismology (Robert Mallett)

bacteria (Christian Gottfried Ehrenberg)

gravity (Isaac Newton)

black hole (John Wheeler)

liquid crystal (Otto Lehmann)

dark energy (Michael Turner)

inflation (Alan Guth)

quark (Murray Gell-Mann, appropriating a word coined by James
　　Joyce in *Finnegans Wake*)

laser (Charles Townes)

scientist (William Whewell)[8]

When I look at this list, I start to wonder if maybe coining words is part of my *job* as a scientist. I've come to think that, as a scientist, if you're not coining new terms or at least struggling to find new terms, you're doing something wrong.

One young clueless scientist I know (me) got his start in science by owning the term *band-limited mask*, together with Wesley Traub. (A band-limited mask is a particular kind of device that you can insert into the beam of a telescope to block the light from a star so you can see faint planets orbiting the star.[9]) The name Wes and I gave it is not the stickiest buzzword ever, but it was indeed a newly made-up word. I think the newness of it caught people's attention and piqued their curiosity. At the time, I felt somewhat forced to dream up a new term to convey that there was a new concept behind the masks—a trick that allowed these masks to block starlight much more thoroughly than their predecessors.[10]

Though the origin of the term felt organic, I got to the point of owning this word in people's minds through a conscious, manufactured effort. Wes and I wrote our paper and passed it around, and a few of our close colleagues got excited about it. Then I felt that I had to fend off other people and prevent them from twisting it, claiming it for themselves, or lumping it together with other ideas. So I wrote a second and a third paper about the masks, each time with different coauthors.[11] I decided that I would show up at every single conference on planet finding and give a talk about the masks. Whenever anybody asked me what I was working on, I told a story about my masks.

They say there are four stages in the acceptance of an idea:

1) Never heard of it.
2) It's wrong.
3) It's probably right, but it's irrelevant.
4) It's what I've been saying all along.

I met all four kinds of opposition. I remember that the fourth stage in particular was hard to challenge with any sort of logical argument: who knows what ideas might have flashed through people's heads? But if anyone had thought of something like it before, they had not published it; I combed through the literature to check. And by the time I'd written three papers on the topic, it didn't matter anyway, because nobody could dispute that I was the current top expert on the topic. Sooner or later, everyone in the relevant subfield of astronomy knew about band-limited masks and associated my name with the new term. It took only a few days to invent the masks. It took a few years of crusading to create the brand.

By the way, you may have noticed that I said that HAT-P-11b was the first planet referred to in a *press release* as a "super-Neptune." That's because it wasn't the first planet ever to be called a super-Neptune in any absolute sense. Josh Winn had referred to another extrasolar planet, HD 149026b, as a "super-Neptune" in the *Astrophysical Journal* two years earlier.[12] And astrophysical theorists had been using the term before that. Of course, none of that earlier use of the term mattered to the public, or to Google. What mattered in this arena was the combination of the relatively new term and the HAT team's bold decision to publicize the heck out of it. That's branding.

Ries and Trout Law #13: You Have to Give Up
Something to Get Something

I took a year off from graduate school to work in Mad Dog Recording Studio in Burbank, California. I found the place in a help-wanted ad in a music magazine, and I didn't know much about it till I arrived for my first day of the internship. The studio turned out to be the musical home of country singer Dwight Yoakum, standard bearer of the California Country sound, known for hits like "Honky Tonk Man" and "Guitars, Cadillacs."

Dwight was and still is a huge country star. Johnny Cash once cited him as his favorite country singer. He played concerts to stadium crowds in skin-tight jeans and a cowboy hat, worn low across his forehead. His public image is tight and consistent, well matched to his songs about hillbillies, old jukeboxes, and the loneliness of men who just can't settle down.

But as I learned, Dwight also has a taste for film noir, jazz, and Woody Allen movies. One night I finished my chores early, and we found ourselves sitting around the recording studio together, waiting for the engineers to wrap something up in the control room—it was just me, the Caltech nerd, and Dwight Yoakum, the Hillbilly Deluxe, watching the second half of some movie on cable. To my shock and amusement, Dwight launched into an erudite critique of the movie, comparing it to Fellini, but finding the color temperature of the lighting unsuitable. I was sure that his analysis would have delighted any one of my visual-studies classmates at Harvard. That's a side of Yoakum that doesn't seem to show in his concerts or albums.

Ries and Trout point out that you can't create an effective brand that sells two very different products, like applesauce and toothpaste. For Yoakum, following this rule seems to have meant suppressing at least one side of himself in his public image. For us scientists, that probably means we have to focus on one subfield of science at a time.

In graduate school, we are often encouraged to try a few different research directions—experiment, theory, and so on—to find out what suits us best. Indeed, that may be what's best for a student's personal growth. But Ries and Trout seem to be telling us that when it comes time to market ourselves, it's best to pick one subject as soon as you can and focus on it until you become the world's expert in it. Once you have gotten the job you want or become the person who always gets invited to speak about this subject at every relevant conference, then you can change directions if you feel like it.

Ries and Trout say that to achieve this kind of focus, you will sometimes have to give things up. This lesson seems to be especially important for postdocs. There are so many things to work on, so many meetings to attend, so many teams to join. But if you try to be everything, you will be nothing.

It's not just postdocs who must strive to focus. Caltech physicist and author Sean Carroll wrote an article for *Discover* magazine's Cosmic Variance blog called "How to Get Tenure at a Major Research University." The article is cynical, but engaging. And among Carroll's top tips for getting tenure, he lists "Don't dabble." He explains, "You might think that, while most of your research work is in area *A*, the fact that you wrote a couple of papers in area *B* will be taken as positive evidence of your breadth and intellectual strength. Very wrong. What will actually happen is that your work in area *B* will be compared to the best people in the world who spend all their time thinking about area *B*, and you will probably come up wanting."[13] In other words, nobody who is shopping for applesauce cares about how well you can make toothpaste.

Ries and Trout Rule #19: Brands Die

The Rubik's Cube, Cabbage Patch Kids, the Spice Girls—all these were effective brands that for various reasons outlived their welcomes. Needless to say, this kind of brand expiration happens in the music business all the time as well. Songs climb the charts and then fall again. Artists come and go. To stay on the charts for more than a few years you have to keep changing, like Madonna or Paul Simon, who have updated their images and musical styles every few years.

Likewise, our scientific brands will inevitably fade. The market may become saturated, for example, once everyone has already read your paper or seen your talk. Or perhaps a better idea will come along. Even if you have discovered a deep truth about the natural world that will live forever in the textbooks, it will one day simply go out of fashion as an idea to study.

When you sense this happening, it is simply time to move on and start working on a new brand. You can be prepared for this eventuality by keeping your finger to the wind and not letting your ego get in the way of moving on to something new. I think it helps to understand yourself. Are you finding work to be less exciting, more stressful than usual? Maybe the brand you were working on has begun to fade. That's when it's time to pop open a bottle of champagne, toast your successes, and then try moving in a new direction.

I hate the word "reinventing." It seems to me that if you invent something it generally stays invented and never needs to be invented again. But when people in business talk about "reinventing" themselves or their brands, they mean they are making a sudden, drastic change of direction so they can have another shot at climbing up the charts. We scientists have to do that, too, every now and then.

How long does the science brand lifecycle usually last? One scientist told me that he reinvents himself every seven years. I like to tell my graduate students that they need to launch one major new brand before they graduate, and then another by the end of their first postdoc.

Brand Names

We talked earlier about the importance of owning a word or phrase. For scientists, that phrase can be the name of your brand—a real word that you can coin. For example, "fractal" and "laser" are made-up words—in this sense, they are no different from "Big Mac" or "Tylenol." Marketing books, such as *Made to Stick* by Chip Heath and Dan Heath, contain what seems to be good advice about to how make up effective brand names like these.[14]

When you're making up a new term, the goal is to create a word that people will like to say and repeat. Here's how this is supposed to work. Two scientists are at a conference. One picks up the program and sees a talk entitled *The Lost Treasure Algorithm*. He can't help but say it aloud. "'Lost treasure algorithm'? What's that?" Someone overhears him and repeats it. Maybe right now you're sitting there wondering what the lost treasure algorithm is.

The phrase "The Lost Treasure Algorithm" works partly because it has two key properties: novelty and mystery. Naturally, a brand name must be novel; if people have heard it before, then they will have nothing to talk about. A brand name should also be somewhat mysterious, or open to interpretation. The idea is to invite the consumer to tell himself a little story about how the name is related to the product. If you make the brand name too literal, there will be nothing left of the story for the customer to piece together.

One trick for creating novelty and mystery is to spell your brand name with letters that are worth many points in Scrabble, such as *z* and *q* (Table 5-1).[15] These letters are uncommon, which helps with the novelty part, and exotic sounding, which helps with the mystery part. As Amy Lo from Northrop Grumman put it, "Anything with the letter X in it is cool by definition."

Table 5-1 High-Scoring Scrabble Letters

Letters	Points
B, C, M, P	3
F, H, V, W, Y	4
K	5
J, X	8
Q, Z	10

Often scientific brand names are acronyms. Acronyms solve the problem of creating mystery and novelty, often adding legitimacy to a name because they can incorporate already-existing terminology. I think brand names based on acronyms are a great way to market your work to colleagues. The only problem with acronyms is that the public finds them off-putting. They will probably be edited out of any press releases you might want to write by press officers looking to protect you from obscurity.

Some companies create a whole suite of products, tied together in name by a prefix or suffix. For example, Apple produces iMacs, iPods, iPhones, and iPads. The same phenomenon occurs in science, though more chaotically. For example, there are many sciency buzzwords with prefixes "hyper-," "super-," "nano-," and "bio-." In astronomy, there was a period when everyone was naming new objects with the suffix "-sar," standing for "StellAR object." That's how we got *pulsars*, *quasars*, and even *blazars* (blazing quasi-stellar objects).

Clearly, the buzzy prefix/suffix approach often works. But a glance at the list of words coined by scientists, above, tells us to keep a wary eye on buzzy prefixes and suffixes; they can quickly become stale and overused. Borrowing a trendy prefix might yield some temporary gain, but history, it seems, is often kinder to those who dig for a completely new term that better differentiates their ideas from the rest. Also, Apple Computers may benefit from having many products with the same prefix, but you might not work on the same team as the other scientists who are using a word prefixed with "nano-." You'll be forced to share some of your glory with other "nano" people if you use that common prefix.

One common complaint I have heard is that some scientists try to coin new terms for something that already exists. Lest people fear that I am trying to warp young minds and sow confusion and grief, let me take this opportunity to say that, at least in science, both the name *and* the idea need to be new. Taking other people's ideas and trying to re-brand them as your own will win you enemies.

Using a Logo

The National Postdoctoral Association (www.nationalpostdoc.org) is an organization founded to help improve the lives of postdocs in the United States. In 2010, the NPA scored a minor coup. They proposed that Congress recognize the week beginning the third Monday in September as National Postdoc Appreciation Week. And just in time, halfway through that week in September 2010, Congress passed H. Res. 1545, officially recognizing the event.

Part of how the NPA achieved this success was by just by spreading the word about National Postdoc Appreciation Week. They sold T-shirts, coffee mugs, and other items to promote the event. The items bore a clever logo with an irresistible, slightly grungy, homemade feel to it (Figure 5-1).

Postdoc and NPA member Jay Morris made the logo using Adobe Photoshop. Zoe Fonseca-Kelly, then chair of the NPA outreach committee, turned the logo into T-shirts. She uploaded the logo into Cafepress.com,

Figure 5-1   A logo used by the National Postdoctoral Association. It's clever, original, and real, and it sold T-shirts. Source: Jay Morris.

a web company that makes custom T-shirts to order. Cafepress charges the cost of making the T-shirt to the person who buys it, plus a royalty on top of that, which it pays to the shirt's designer. So Zoe and Jay were able to make the T-shirts at no cost to the NPA and sell them online to customers with no initial investment. "We did everything in Cafepress.com," Fonseca-Kelly told me. "It's perfect if you can't afford to buy hundreds of T-shirts yourself. They provide you hundreds of items you want, from Christmas ornaments to mousepads. You just upload your logos onto it and then you get a T-shirt! The disadvantage is that it's a little expensive. But you don't have to outlay anything personally for it." Zazzle.com provides a similar service.

Using a logo is an essential way to show you mean business, and to make your brand more memorable and recognizable. We are all competing for space in the minds of our customers. But when you add a logo to your branding scheme, you tap into another portion of your customer's brain—their visual memory—that they can use to remember your brand.

Unfortunately, just as inventing a new word sometimes comes with a temporary stigma for scientists, so does designing a logo. But I hate to see young scientists scared away from this career-building tool. I think the time arrives for a scientist to design a logo when he or she begins to form some kind of scientific team. A lone scientist trying to use a logo might be viewed as self-promoting. But there's nobody to blame if the logo comes from a team of scientists, and every team can benefit from using one. As Adam Burgasser, professor of physics at University of California San Diego, told me, "The logo provides a nice visual cue for the scope of the project and helps gather everything associated with it under one tent." Maybe scientific teams are a little bit like sports teams; a logo provides the face of the team's collective identity. So I say: if there's a team, there should be a logo.

The scientists on the Marketing for Scientists Facebook group told me a little bit about how they make logos. Often, scientific logos seem to start out as key figures (illustrations) from some paper; sometimes you can make a figure into a logo by making the lines thicker, or zooming in on a recognizable patch of the image. Ideally, a logo is still recognizable even when you mangle it a bit—by printing it small or in grayscale, or by viewing it from far away.

Obviously, logo design is a craft and also a big business, and some scientists hire professional graphic artists to make their logos. Adam Burgasser said, "I simply put an ad out on Craigslist for a graphic artist to make me one on spec. I got many replies, picked the artist whose work I thought best matched the idea and my taste, sort of sketched out my idea, and ended up with a graphic I was very happy with and use often (and it was only $500)."

But you can design a logo yourself if you have a bit of an artistic eye, and a deep fear of clichés. In fact, if you announce to your team that you need a logo, you might be offered ten logo ideas in a blink of an eye. You may even want to hire a professional artist just to avoid having to resolve an argument among your team about whose logo idea is best.

I asked Jay Morris how he made the NPA logo, and he said, "I made the knife and fork and sleeping guy myself, just using the colored fill-in shapes in Adobe Photoshop. I made a rectangle and made an oval, then cut away part of the oval to make the tines. I knew I didn't have to get really detailed because I knew we didn't need much resolution." Curiously, Morris also told me that the Google search engine is one of his favorite design tools. For example, you can just type "christmas tree in photoshop" into Google, and it will give you a recipe for drawing one. Jay also mentioned the online resources at Adobe.com and Digital-photography-school.com.

Now let's talk about what makes a good, memorable logo. A classic way to design a memorable logo is to build into it an optical trick, a little puzzle that draws the eye back in for a second look. Here are a few well-worn tricks that probably come to your mind when you think about logos:

Letter substitution: Write down the name of your brand, then remove one of the letters from the name and replace it with a different shape or image.

Word art: With a little stretching and fussing, you can probably arrange the text in your brand name into a little picture.

Knocked-out text: Make the text white, like the background, and lay it on top of a solid, dark shape.

These optical tricks were meant to help make a logo memorable. And they used to work well before they became clichéd. You might still be able to get away with them if you apply them creatively. But I want to point you in a different direction. It's more important to be different and original, even at the cost of a polished look; for example, a simple doodle can be a good logo as long as it's original.

The qualities that make a good science logo are the same qualities that make a good brand name: novelty and mystery. A logo has to be new; it has to be unique to your brand. And it should ideally leave something to the imagination and should inspire people who see it to tell themselves a story about how it is related to your brand.

Let's talk some more about novelty. As we were discussing, most of the time, when scientists try to design logos, they reach for standard tricks like the ones I mentioned above. Or they grab some stock clip art from the Web. The result is often something that looks like you've seen it somewhere else before. For example, Figure 5-2 shows a fake logo that I made using a free logo template. It certainly looks logo-ish, but it's clichéd and forgettable. That's something to avoid.

Also, scientists making logos are often tempted to mimic the themes in the scientific logos we see around us, like the ubiquitous atom with the electrons orbiting it, or the double helix, or the clichéd corporate swoosh. That's another path to obscurity. Part of achieving novelty is recognizing the design elements that scientists will find clichéd. Scientists have already overdosed on test tubes and rockets and hexagons and globes and orbiting things and molecules and green plants. Instead, how about a red medieval castle? A blue pot of coffee? A multi-colored cat?

National Postdoctoral Association

Figure 5-2   Fake logo for the National Postdoctoral Association made using a free logo template. It certainly looks like a logo, but it's clichéd. You would not be excited to wear this on your T-shirt. Source: Pete Yezukevich.

I recommend reading the yearly trends articles at LogoLounge (www. logolounge.com/), which address the need for novelty head-on. They discuss the latest clichés in logo software and logo design—to help you avoid them.

The element of mystery can be just as hard to conjure as the element of novelty. But it's necessary so that people will project their own meaning into your logo, just the way they might project meaning into a brand name or, say, the lyrics of a song. Here is one example of a logo that succeeds in this regard: Freedom Scientific's logo is a hot-air balloon. Note that Freedom Scientific does not make hot-air balloons. They make products for the blind and visually impaired, like Braille displays. The logo forces you to discover for yourself the connection between Braille displays and the balloon. The hot-air balloon, you realize, symbolizes the feeling of elation a blind person might have on using a product that allows him or her to read. When you discover this connection on your own it becomes partly your idea, and it sticks in your mind better than it would if it were purely someone else's idea.

Figure 5-1 shows Jay and Zoe's National Postdoctoral Association logo. This logo is a parody, one of many "Eat, Sleep" logos that celebrate sports and other intense activities. (You can find T-shirts that say "Eat Sleep Hunt," "Eat Sleep Sail," "Eat Sleep Hike," etc.) With its bold graphics that imitate a road sign, it lampoons the very idea of a logo—a snarky, iconoclastic gesture appropriate for an organization of young PhDs fighting the powers that be. "Who came up with such a fun concept?" you find yourself thinking. Then, you answer your own question: smart people my age whose association I'd like to join.

Passion

One summer I went on a trip to France, a great place to learn about passion. I was there for almost two weeks, during which I often found it necessary to probe the chemistry of the local *chocolat chaud* or test the structural integrity of the local croissant. So during that time I probably sat in twelve different cafés in different corners of Paris, Versailles, and Chartres.

Most of the cafés were individually forgettable, but there was one particular café that stands out in my memory. This one had, according to a sign by the door, a collection of 357 teapots. There were Russian teapots,

Japanese teapots, teapots made of ceramic, bronze, and bone. While I was there, I just had a cup of coffee, and it was about the same as any of the many cups of coffee I drank on that trip. But of all the cafés I went to, this is the one I remember most, with its 357 teapots.

Thought it's not exactly that the teapots themselves enhanced my meal. They weren't being used, except for decoration, and I didn't even drink tea. And though the teapots added to the décor, it was not in proportion to their numbers. I might have studied two or three teapots that were displayed near my seat. Then I glanced around the room where I sat and took in the visual impression of maybe sixty or seventy more.

What stuck in my mind was the thought that somebody behind the scenes in that café really, really liked teapots. My grandmother liked teapots, too. But that meant she had six of them. Whoever was behind the teapot collection in that café in Chartres liked teapots about sixty times as much as my grandmother did. That's an extreme teapot fetish.

I mentioned above how the idea of branding came from the cattle industry; a brand is a way to distinguish your cow from the rest. But what if your cow were intrinsically different from the rest of the herd, as this café with all the teapots was different from the other cafés in France? Marketing guru Seth Godin wrote a book about this concept, which he called *purple cow*, borrowing a phrase from a short poem by Gelett Burgess.[16] A purple cow can be any kind of brand: a person or a company or a product. In a way, it's best if it's all of the above—a complete package.

Godin argues that a natural way to form a purple cow company is to start with a purple cow person, someone with a deep personal obsession—a passion for some specialty. When people realize just how passionate you are, they can't help remembering it and talking about it. That passion, and that brand, can translate into an intrinsically outstanding company, like the café with the teapots.

Passion for their research is something many scientists have naturally. It helps that we often find subjects more important than teapots to focus our attention on. Take ecologist Cristina Eisenberg, director of research at the High Lonesome Ranch in Colorado—a richly diverse chunk of conservation land comprising nearly 300 square miles. Eisenberg is passionate about wolves.[17] When the wolves disappear from an aspen forest, the elk in the forest can become overpopulated and start eating the aspen trees. Eisenberg's passion has driven her into the forest to investigate thousands of pieces of wolf scat and measure the diameters and heights of thousands of aspen trees. Her efforts help rescue both wolves and irreplaceable forest.

"Research and being a scientist is a lot of work," Eisenberg told me. "Some people think it is very tedious. Why do I run around measuring thousands of aspen trees looking to see if they have been eaten? It's because those trees tell a story and those stories cover a whole ecosystem." On the one hand, you might say the aspen measurements are just data points. On the other hand, you might think of them as testimony to Eisenberg's passion and dedication, irresistible components of her brand. The café in Paris had many, many teapots; Eisenberg has many, many aspen measurements. The teapots helped the café make a lasting impression on me, and the aspen measurements helped Eisenberg win millions of dollars in grant money.

Thinking Big

Individual scientists can be remarkable for their dedication and passion. And a science project can be a purple cow too, expressing, in people's minds, the dedication and passion of the scientists behind it. It seems to help if a science project is vast in scope. For example, the Human Genome Project, when it was first imagined, seemed so dauntingly huge that even before it was funded it was newsworthy, just for the audacity of the idea. That's the real secret of the marketing campaign for the project run by James Watson and Francis Collins and others: the product was so outrageously tremendous that people had to talk about it.

The Living Earth Simulator recently proposed by Dirk Helbing at the Swiss Federal Institute of Technology in Zurich is a new big science project that's been in the news lately. It's a proposed giant computer model that simultaneously incorporates everything we know about the planet Earth and its inhabitants, from climate to sociology.[18] The computers of today are too slow to handle it, but Helbing argues that by the time the software and databases are ready, the computer technology will have caught up. You can't help but remark about that!

I would like to point out that Helbing's website photo shows an image that's very different from the image Watson tried to portray when he was canvassing for funds, in the anecdote in chapter 3. Helbing wears a crisp suit and tie in his photo—there are no untied shoelaces. Perhaps Helbing is trying to a project a different kind of brand than Watson, one corresponding to a different societal role—the role of the leader, the professional. (We'll talk more about this idea in the next chapter.)

Branding Is a Painful Process

I was delighted to read in *The Big Book of Marketing* that emotions can ride high in a branding project. I was glad to see that someone else recognizes this fact. It's true for businesses, and I know from experience that it's true for us scientists. If you are working by yourself, the emotional stress of branding might not matter. But if you are working with a team, it surely will—especially if it's a team of scientists.

As you know, scientists are creative people with fragile egos. It can often be hard to get scientists to answer their e-mails. But when it comes time to choose a name for whatever project you're working on, sometimes everyone wants to weigh in, and nobody seems quite happy with the outcome.

I hope you won't let the pain of this process stop you from working on your brand. Sooner or later, your brand will become accepted by those around you. You'll be shocked to hear your brand name mentioned in casual conversation as though it has always existed—and you'll know you're on the right track.

A related problem is that people around you can get nervous when they sense a change in *you*, like when you decide to launch a brand, assemble a team, and lead a project. It's just like in high school, when one day I realized I wanted to wear cargo pants (don't laugh; they were in at the time). My friends started getting jealous and telling me that I had changed. Colleagues can get jealous when you change and grow—and I often find that I need to go out of my way to show them some extra attention and reassure them. But as you know, that's no reason not to change and grow.

A Battle for Space in the Consumer's Mind

As Ries and Trout describe in their books and other writings, marketers often view branding as a battle for space in the consumer's mind. Consumers, they recognized, are steadily bombarded from every direction with advertisements and other messages—too many to believe or process. So consumers must steadily filter these messages, throw away junk mail, delete spam, and so on. By and large, the job of a marketer is to figure out how to get past the filters, by making their brands exciting, yet also comfortable, familiar, and consistent with the consumer's prior experiences.

It seems to me that the same picture holds in science as well. Science is a kind of battlefield of ideas and brands. We come up with sticky new ideas and we do our best to spread them around and develop their reputations (brands). We recognize that our colleagues are all steadily bombarded with information, and we seek to make our ideas attention-worthy.

We scientists must learn to love the battlefield of ideas, memes, and brands—we often view ourselves as proud warriors on this battlefield. As a scientist, I feel like I am perpetually running around, sword in hand, defending some scientific result or another. It's become part of my personality that everyone around me has come to expect.

Indeed, I like to argue that with our warrior's stance, many of us scientists have chosen a particular role for ourselves in society—one that sometimes helps us, and sometimes hurts us. It's a role and a perception that we would do well to understand if we want to tell our stories and further our causes. And that brings us to a new topic, one that I find marvellous and mystical: how to brand ourselves through archetypes.

CHAPTER SIX

Archetypes

 As I have already mentioned, people love stories. More than that—people *live* stories. We humans tend to subconsciously picture ourselves as living, walking characters from a book or movie. That's more or less what this chapter is about: the sensation we sometimes have of living our lives in a story, and being the characters in our life stories.

I'm going to tell you about a concept called "archetypes" invented by Carl Jung in roughly 1919. I find this notion magical, a key to understanding country songs—and my own life. I've come to believe that it applies to aspects of science marketing as well.

For example, have you noticed that some young scientists just seem to radiate success as they race straight down the path to stardom right from the start? Their advisors and teachers speak adoringly of them and write them glowing letters of recommendation, overlooking their flaws. This chapter may explain something about this phenomenon.

Twelve Stock Characters

Archetypes are the stock characters in legend, in movies, in the stories of our lives. Some kinds of characters are so fundamental to human experience that they need no explanation. The moment they walk onto the set, we know who they are.

Maybe the best way for me to illustrate the concept is by example. Consider the characters in *Star Wars IV: A New Hope*. Luke Skywalker, dressed in white, begins the movie as an innocent. Grey-bearded Obi Wan Kenobi is a sage. Han Solo is an outlaw. Darth Vader is a kind of evil ruler.

The theory of archetypes says that the innocent, the sage, the outlaw, and the ruler appear in many movies, novels, legends, and stories of all kinds, from the Bible to Harry Potter.

In their book *The Hero and the Outlaw: Building Extraordinary Brands Through Archetypes*, Carol Pearson and Margaret Mark claim that brands fitting one specific archetype, rather than a blend of archetypes, fare better in the marketplace. Many companies have capitalized on this advice. I'd like to argue that scientists could benefit from it too.

Pearson and Mark identify eight more archetypes, besides the innocent, the sage, the outlaw, and the ruler—twelve key archetypes in total. There are other systems of archetypes, but this one seems to be the most popular in the business world. Here is the complete Pearson and Mark list, together with some examples of characters and brands that exemplify them:

> **The Innocent:** The innocent feels that if you wait, everything will be just fine. The innocent could be a child whose parents provide for him, who remains untarnished by exposure to the struggles of life. Or it could be an adult who has come to understand that what will be will be, and has found peace with the world. Forrest Gump appealed to viewers not for his intelligence, but for his innocence. Ivory Soap is 99.44 percent pure. Coke has a naïve plan to reach world peace: it would like to teach the world to sing.

> **The Sage:** The sage dispenses advice. This is a typical archetype for professors. The sage can be a regular pedant or a know-it-all. But there are hip, modern sages too, like Oprah and Martha Stewart.

> **The Outlaw:** Outlaws can be good or bad, but either way, they hate authority and the status quo. If you want to feel like a rebel, you can don a Harley Davidson leather jacket or fly on Virgin airlines. John McEnroe. Anthony Bourdain.

> **The Ruler:** The king, the chief, the president, the boss. When you use your American Express card, you'll feel like you're in charge. IBM, Microsoft, Frank Sinatra, Rolex. The Rolex watch company even has a crown as its logo. Darth Vader is a kind of evil ruler.

The Hero/Warrior: The hero is the archetype that faces challenges and takes responsibility. Superman. Sally Ride. Put on your Nike sneakers, for example, and you'll be ready to win the Olympics.

The Explorer: Brands like Levi's and Jeep make you feel ready to explore, ready for adventure. Indiana Jones. U2's song, "Where the Streets Have No Name."

The Wizard: The wizard creates real change—seemingly out of nothing. Oil of Olay, for example, performs the magic of "age defying" regeneration. The wizard is a good archetype for a management consultant, who may be called in to magically rescue a company from doom.

The Lover: The lover dedicates himself or herself to one other person alone. Victoria's Secret is for the lover in you. Ditto for Barry White.

The Everyman: The everyman is someone who is proud to be a regular guy, an unpretentious man of the people. The everywoman is the girl next door. If it is possible to be proud to be humble (see chapter 3), then this is the character that can pull it off. Budweiser. Ford. Bruce Springsteen. I read somewhere that Mike's Hard Lemonade was named "Mike's" simply because Mike is such a common male name.

The Joker: Jokers speak the truth, the harsh truth, the beautiful truth—which sometimes makes people uncomfortable, sometimes puts them at ease, and often makes them laugh. Conan O'Brien. Robin Williams. Alka-Seltzer. Plop plop, fizz fizz, what a silly song that is!

The Caregiver: Everyone has an instinct to be nurturing sometimes. And if you really care about someone, you'll send a Hallmark card, and drive her around in a Volvo.

The Creator: With Legos, you can build anything you can imagine. Home Depot. The Blue Man group. Apple. Google. TCHO.

Most of these characters appear regularly in country songs (though the creator and wizard seem to be pretty rare). In fact, you might say that hit country songs can often be construed as sales pitches for a particular archetype.[1] One obvious example is the love song. The verses of a love

Figure 6-1   Archetypes: the creator, the ruler, the hero, the everyman, and the innocent. Source: Pete Yezukevich.

song might explain how the singer was sad and confused. Then the chorus reveals the solution to his woes: love, true love. The character singing the song is an archetypal lover because now he focuses solely on one woman (at least till the end of the song).

I like to say there are eleven other kinds of songs besides love songs. One common kind of country song is the everyman / everywoman song, a song that celebrates how good it is to be a regular guy or regular gal. One example of an everyman / everywoman song is "Redneck Woman" by John Rich and Gretchen Wilson. It's about a woman who's proud to be who she is, though she doesn't have fancy clothes and she prefers beer to champagne. Another common kind of country song is the hero song, which celebrates someone facing adversity. Country singers face plenty of adversity. I once wrote a hero song called "I Can Break My Own Heart," where the singer runs into her annoying ex who dumped her, and she tells him to shove it.

Besides love / lover songs and everyman / everywoman songs, there are joker songs, innocent songs, wizard songs, ruler songs, and so on. Each one of these twelve categories of songs resonates with us by tapping into a different one of the stock characters we have in our heads. It's not just country music either; pop, rock, R&B, hip-hop—every kind of popular music hews to these archetypes.

Archetypes and Scientists

I'd like to suggest that scientists, consciously or not, incorporate archetypes into their brands, too. For example, Dirk Helbing, originator of the Living Earth Simulator we talked about in the last chapter, wears a crisp suit and tie in the photo on his homepage, thus evoking the ruler archetype. Here are some more examples:

Carl Sagan: Innocent

Carl Sagan's television shows emphasized humankind's small place in the universe, the connectedness of everything, the miracles of the natural world. This attitude of wonder lent an innocent appeal to Sagan's popular work. Sagan said, "Our species is young, and curious, and brave. It shows much promise."[2]

Stephen Hawking: Hero

Crippled by amyotrophic lateral sclerosis and unable to speak, Stephen Hawking communicates with the world using a special device that allows him to type onto a computer with small movements of his cheek, one letter at a time. Nonetheless, Hawking has managed to write many influential scientific papers and books and to become a major figure in theoretical astrophysics and cosmology. This inspiring story of triumph and endurance has made Hawking a hero in the public eye.

Jane Goodall: Caregiver and Hero

Jane Goodall is known for her groundbreaking work on the tool-making habits of chimpanzees—and even better known for taking up the challenge of protecting chimpanzees and their habitat. She projects the image of a caregiver and a hero, swooping in to rescue these delicate creatures from the predations of humankind.

Richard Feynman: Joker

Scientists may know Feynman for helping formulate the theory of quantum electrodynamics. But Feynman is probably best known to the public through his book *Surely You're Joking, Mr. Feynman*. The book tells stories of his pranks and practical jokes and how he outwitted various competitors, laughing all the way.

Albert Einstein: Creator

"Imagination is more important than knowledge." "Anyone who has never made a mistake has never tried anything new." Einstein's well-known quotes celebrate imagination and creativity. Famously childlike, Einstein had a reputation for treating modern physics as a kind of playground. His discoveries provided the tools for others to build the photoelectric cell, the atomic bomb, and many other creations. One might even say his theories underlie the creation of the universe.

Richard Dawkins: Outlaw

Dawkins is known for making provocative statements applying the theory of evolution to the development of religion, among other things. His website, RichardDawkins.net, fuels his rebellious image. The title at the top of the page says "A Clear-Thinking Oasis," implying that Dawkins is proudly defying the mainstream. Further down the page, last time I looked, were some images promoting a Dawkins DVD, flaunting the fact that the DVD was "banned in Turkey."

It appears that many different archetypes can work for scientists. The hero / warrior is one I mentioned earlier. But I know successful scientist wizards, outlaws, explorers, creators, rulers, and even jokers. If you are a student scientist, perhaps you are projecting the image of the wide-eyed innocent.

Science projects, not just individual scientists, can embody archetypes, too. Here are two salient examples from my field:

Hubble Space Telescope: Hero

The Hubble Space Telescope was "born" with a physical deformity, a few millimeters of optical aberration that rendered its images blurry. But a rescue team of astronauts soon repaired the problem, and the Hubble became a hero for its perseverance and triumph over adversity. The triumph yielded a well-known treasure trove of astronomical observations.

The Mars Rovers: Innocent

With their big electronic "eyes" and their childlike crawl, the Mars rovers "Spirit" and "Opportunity" captured the hearts of America the same way a baby panda bear could. Soon their likenesses became popular children's toys. It seems a shame these rovers were given such explorer names; to match their archetype maybe they should have been named something like "Tigger" and "Pookie."

Sometimes I've been asked, What is your archetype? It's not an easy question to answer. The theory says that we each contain all of them. Maybe at home you are a caregiver. With your friends, maybe you are the everyman. At work, maybe you are a hero/warrior. It is wonderful and natural that you have all these characters inside you. And it's fine to choose one of them at a time to live and project.

Carol Pearson argues that to achieve happiness a person must live out and deeply experience each one of the various archetypes. But sometimes an archetype can lie dormant inside us. An archetype depicted in a movie or a song helps people witness the full range of human potential; it can help us unlock hidden aspects of ourselves. That's part of how the right song playing on the radio at the right time can pick you up when you're down.

The same power of connection applies to brands that use archetypes, like the ones I mentioned above. It explains why people don't just like the taste of Coke, for instance. They love it like they love the familiar innocence of a teddy bear.

But while all the archetypes have something to offer us as human beings, I think that some archetypes historically do not work for the public images of professional scientists. I was unable to think of any famous scientists who benefitted from associating with the lover archetype, for example.[3] This archetype may work well for singers of popular music, like the R&B group BoyzIIMen, but it seems to me like you are taking a risk trying to make the lover archetype work for you as a scientist. If you find yourself wearing lots of jewelry and have your shirt open at scientific meetings, some people might view you as foolishly vain and unfamiliar with the culture of science. Agnes Kim, assistant professor of physics at Georgia College suggested, "The key is to look professional. The sexier the look, the more professional it needs to be."

Of course, since you are a scientist, I have no doubt that you are a wonderful lover. Let me emphasize that there is no need for you to turn off this part of your true self. But when it comes to managing your professional image and your personal brand, you might want to choose a different aspect of your many-faceted personality to highlight.

Another example of a potentially dangerous archetype seems to be the caregiver. In *Nice Girls Don't Get the Corner Office*, Lois Frankel advises, "Unless you're Betty Crocker, there shouldn't be home-baked cookies, M&Ms, jelly beans, or any other food on your desk. . . . We don't ascribe a sense of impact or import to people who feed others."[4] In other words, Frankel is warning women that they should be wary of projecting the caregiver archetype at work. There are exceptions to this rule, of course, and Frankel is quick to point out that a bowl of candy on your desk can help you in situations where you need to soften your image, or make people comfortable; the key is to be aware of what you are communicating. By the way, *Nice Girls Don't Get the Corner Office* is one of my favorite marketing books. It offers 101 marketing tips, and only a few of them apply exclusively to women.

I think that the hero/warrior archetype is especially important for junior scientists to keep in mind. When you hear the word *warrior*, you might picture Rambo, covered with gun belts, or an ancient Greek gladiator. But there are also warriors in the professional world, wielding pens and briefcases. Warriors might not always carry swords, but they are always competing and trying to be the best.

The language of recommendation letters is the language of the hero/warrior. Who is the best young scientist? Who has the most publications? Letter-writers struggle to pack their letters with comparisons and superlatives, evoking the warrior's fight. My suggestion is that if you can evoke the hero/warrior archetype, you might be able to send a subconscious message that you are a winner. If you send this message to someone who is writing you a letter of recommendation, it might help her describe you in the right language.

I was once talking to a friend of mine, Princeton professor of Mechanical and Aerospace Engineering Jeremy Kasdin, about his graduate student. He said, proudly, that the student was almost ready to graduate; "He's even starting to look like a scientist." I think that part of what he meant by this is that the student was leaving behind the innocent archetype of the student, and taking on the characteristics of a hero/warrior.

Famous female scientists—Jane Goodall, Rachel Carson, Sally Ride—often seem to be heroes. Maybe that's simply because succeeding as a female scientist makes you a hero, triumphing over the adversities women face in a male-dominated field.

Another good archetype for a scientist is the ruler. Leadership is a much sought-after quality in a scientist, especially a scientist-manager. But it is hard to judge scientific leadership, just as it is hard to compare two young scientists and decide who is "better." So I suggest that if you are looking for a letter of recommendation that emphasizes your leadership, you might want to help your letter writer by looking the part, possibly by dressing a little bit more like the rulers in the American Express ads. Pinstripes might seem a bit much. But maybe an outfit that emphasizes your height or shoulders would do the job.

Likewise, creativity is a characteristic often touted in letters of recommendation. That makes the creator archetype an important one for young scientists—if you find that it suits you. The black turtleneck, jeans, and short-cropped hair of Steve Jobs seem to speak of a kind of creator (or perhaps a wizard). I have seen many scientists try to evoke that style. Of course, a real creator would probably not just copy someone else's style, but add something creative to it.

Clothing isn't the only way to project an archetype. The ways you express yourself and carry yourself probably contain cues as well. The decorations in your office, the pictures on your website, the images and language in your talks and papers can all communicate archetypes.

The Orphan

Person and Mark say that every archetype has a shadow side. The shadow side of an archetype carries negative qualities. For example, Darth Vader is a demagogue, the shadow of the ruler. The warrior is the shadow side of the hero—more focused on the fight for the sake of a fight. You won't often see these shadow sides in TV commercials; companies don't want to risk creating a negative image for their products. But you'll find these shadow archetypes in entertainment and art of all kinds, like movies, novels, poems, and songs.

The shadow side of the innocent is the orphan. For us scientists, I think this shadow archetype is an especially important one to be aware of. Now, I'm not talking about the outlaw, the archetype of someone who challenges authority. The orphan is not up to making such an effort. The orphan feels like his life is not under his control, so all he can do is whine and complain. The orphan is associated with passive aggression and immaturity. Modern rock songs, where the singer complains about his parents, his girlfriend, and so on, but seems somehow unable to take responsibility for his problems and his life, often lean heavily on the orphan archetype.

I'm sure you've met colleagues who often seem to play the orphan. I'm sure it doesn't serve them well. When scientists take on this character, they signal that they are not ready for responsibilities, like the responsibility of managing a grant or teaching students.

It is easy to feel like an orphan when you are a scientist. Your college, graduate school, or postdoctoral experience may leave you feeling abandoned. Your struggle to find a job may leave you feeling powerless. I have played the orphan myself; it is something we all go through from time to time.

But some people get stuck in a rut and find themselves playing the orphan all the time. That's no fun for you or for the people around you. Carol Pearson offers some advice for people who get stuck playing the orphan.[5] It's hard to break free of this mold on your own; seek out someone who can help you get unstuck. That person could be a friend, a colleague, a doctor, a significant other—anybody with time to really listen or maybe give you a push in a new direction. I know this sounds like a cliché, a soap opera, or a country song. But sometimes life really is like a country song, even for scientists.

CHAPTER SEVEN

The Consumers of Science

Science is a busy marketplace, filled with purveyors of all kinds of products. We participate in the market when we craft papers with the needs of the reader in mind (chapter 2) or when we lobby congress, carrying a prop (chapter 3). We feed it when we coin or use buzzwords (chapter 5) or read or write letters of recommendation (chapter 6). The marketplace hummed when scientists gathered in Versailles, showing off their latest experiments to the Sun King and his court, hoping to score royal capital to finance their research. And we see it going strong at conferences (chapter 4) and at science fairs. But sometimes, surrounded by the cobwebbed pillars of our solemn institutions, we can lose sight of the wonderful marketplace we work in.

So let us use our understanding of marketing to clear some more of these cobwebs for a better view. In the first half of the book, we looked at the principles of marketing and sales, and then tried to apply them to science. Now let's work in the other direction: start with the institutions of science, and see how we can apply the principles of marketing to understanding them. This introspection might prove painful at times, but I believe it will ultimately improve how we do science and how we treat each other.

First, a Sales Pitch

With this book, I am trying to sell you on the use of good marketing practices and on the idea that science and marketing go hand in hand. A good sales pitch will acknowledge your preconceptions of the product being

Table 7-1　How My Understanding of Marketing Has Evolved

My Preconceptions of Marketing	Good Marketing Practice
Giving someone a canned elevator speech	Intriguing someone with a cool prop, and starting a conversation
Sucking up to someone, "networking"	Slowly building a real working relationship
Giving a talk about how great you are	Giving a talk about a new idea that can help other people
Telling your colleagues to "check out your website"	Putting handy software and data on your website that people can use for free
Hunting down important people and monopolizing them	Recognizing when someone feels new and lost, and showing him or her around
Telling someone a story about why you are the best	Teaching and entertaining people with a story they can relate to
Using psychological tricks to manipulate people	Trying to better understand people and connect with them
Trying to change people	Trying to help people

sold and then help distinguish the actual product from those preconceptions. So let me take a moment and contrast some of the preconceptions I once had about science marketing with the modern picture of marketing that I am trying to paint (see Table 7-1). This table will also serve as a handy review of some of the key points of the last few chapters.

With so many bad preconceptions out there, it makes sense that many scientists are put off by even the mere mention of the word "marketing." In *Don't Be Such a Scientist*, Randy Olson writes that roughly one-third of scientists across all disciplines tend to behave in an outreach-phobic way. He estimates the value of this constant fraction to be one-third because that's the fraction of the National Academy that voted against Carl Sagan's Academy membership. I've noticed that there seems to be a similar fraction of scientists who blanch when I mention the word "marketing."

I had a minor revelation about this topic the other day while I was wiping the grease from a giant burrito off of my chin. Many people, scientists included, confuse "marketing" with what is now called "traditional marketing" or perhaps with "advertising," the most salient feature of a traditional marketing campaign. Scientists hate advertising. But we are not alone in hating advertising; that's something we have in common with the founders of Chipotle Mexican Grill.

At the restaurant chain Chipotle, customers design their own burritos, choosing the salsa, the beans, the cheese, and so on. Then they sit down to eat them in a room with hip, industrial décor, featuring stainless steel and corrugated sheet metal. It's meant to evoke a kind of a burrito factory or laboratory—where you are the star inventor. That's the creator archetype, loud and clear.

These days, Chipotle advertisements are commonplace: commercials, billboards. But at first, Chipotle considered itself too cool to advertise. A 2007 article about Chipotle and its marketing scheme from *Bloomberg Businessweek* said, "The Denver-based company eschews TV commercials and most other traditional advertising. In fact, it spends less in a year on advertising than McDonald's Corp., its former parent, spends in 48 hours. 'Advertising,' declares M. Steven Ells, Chipotle's founder and chief executive, 'is not believable.'"[1]

The company's strategy has shifted since then, and now there are billboards advertising Chipotle all around the country. But the ads still tend to be of the too-cool-for-this-billboard variety, featuring colorless pictures of burritos with the aluminum wrappers left completely on, offering no visual hint about the spicy, steaming food inside. They are a far cry from McDonald's billboards, with their glossy images of precisely overflowing hamburgers.

I think that some scientists are marketing themselves *à la* Chipotle circa 2007. They scoff at press releases and public outreach, on the premise that such promotional tools are unbelievable—like advertisements. They hope their products and customer experience will speak for themselves, without a fancy wrapper, and that word of mouth will suffice to spread their ideas.

Of course, Chipotle has always been marketing—whether or not they use billboards or commercials for promotion. The marketing lives in the decisions to serve delicious food that people crave, to serve it quickly so people don't have to wait, and to build clean restaurants with hip industrial décor that people find novel and compelling. Choosing a slogan is marketing, even if the slogan is—like Chipotle's—simply "Food with Integrity." Chipotle has put a great deal of thought into the needs of their customers, and the stories that bring meaning to their customer's lives.

Likewise, even those scientists who squirm away from the idea of it must nevertheless be marketing—deciding what their colleagues and funding agencies want to hear, thinking of how to maintain their images. Maybe a staunch resistance to certain marketing tools is even part of the subtle story these scientists tell their customers. "My work speaks for itself"—that's a kind of slogan, too. Advertisements may indeed be inherently unbelieveable and tacky. But even if you decide not to advertise, you are always marketing, or you don't stay in business.

The Needs of the Scientific Community

Now we are ready to move on and try to apply the principles of marketing to science from the bottom up. We should probably start by thinking about the needs of the people in the scientific community—our customers. That's what we talked about in chapter 2. So I've tried to compile a list of some of the different kinds of people in the scientific community, broken down crudely according to their primary concerns. These oversimplified descriptions may seem like caricatures, but maybe they are a place to start.

CONSUMER: Students

MAIN CONCERN: "What am I going to do with my life?"
Students often make up a large fraction of any academic department—and the audience for your professional talks. At some point, we scientists often want to hire students or attract them to our programs, so it's useful to be able to keep their needs in mind. Here's my take on the needs of science graduate students; please forgive me if you're a student and you disagree, or find yourself caricatured or oversimplified.

Students are always looking for something important and exciting to work on. They love the rush that comes from being around smart people, but they hate to be talked down to. To get a student's attention, show him or her something dazzling and smart, or something that people seem to be talking about in the news. But keep it real. Students are especially sensitive to insincerity; they love to spot a fake and call him on it.

CONSUMER: Junior scientists

MAIN CONCERN: "How can I get a permanent job that I want?"
By junior scientist, I mean someone who doesn't yet have a permanent job, such as a postdoc. A junior scientist is often bloodthirsty, a little bit jaded, and looking for a way to stand out from the crowd. This hunger can make him high-strung and nervous. But, as everyone knows, scientists often make their most creative contributions when they are at this stage of their careers.

One proven way to get a junior scientist's attention is to mention the jobs and fellowship opportunities in your department. You might mention how you need a leader for a new project or a speaker for your conference. And of course, junior scientists always like to know about new scientific tools that they can use and new questions they can answer.

This discussion is beginning to sound like something you might see in a horoscope or hear from a carnival mind reader. Perhaps that's because there's something in common between marketing and mind reading. You have to try to guess what people are thinking, what they want, what's in it for them.

CONSUMER: Senior scientists

MAIN CONCERN: "How can I win my next big grant?"
Senior scientists already have permanent jobs. They have already traveled the world. They are building empires and climbing the ladder in their home institutions, while setting aside time to spend with their families. They want money to support their students and postdocs. They need to find more good people to hire. They are afraid of being overtaken by younger scientists. They are already committed to several different projects and have little time to spare.

To get a senior scientist's attention, help her by offering to do some work, like analyzing her data. Or add her as an author to a paper or a grant proposal. Suggest a good person for her to hire. Offer her a new figure or paper that supports her favorite idea / mission / grant proposal.

CONSUMER: Staff at funding agencies

MAIN CONCERN: "How can I have an impact despite all the pressures and constraints?"

Funding agencies are often staffed by senior scientists who leave universities and labs seeking new experiences, wanting to find ways to affect the way science is practiced on a large scale. I asked Rita Colwell, former director of the National Science Foundation, about what it's like to work for a funding agency. She said,

> When I was at NSF, I was never the last person to leave at the end of each day—there was always somebody there working hard. The motivation was that they love science and they were able to get a 30,000 foot view of what was going on in the world of science and engineering. They make friends around the country and the world while they are in their position at NSF and learn how the science funding system works.

But when a senior scientist leaves her relatively cozy academic institution to work at a funding agency, she might be shocked by the stress of the new environment and all the financial and political constraints. It's hard to make a difference. It's hard enough just to stay afloat. Also, some staff at funding agencies work there for stints of just a year or so; they may try to maintain research programs at their home institutions while they are away. That can be a further source of stress.

Colwell advised me that if you're funded by the NSF you should "send a reprint [of your latest paper] to your program manager so he or she will know that his or her program has resulted in a success. The story goes on the web and it helps NSF's Office of Legislative and Public Affairs release information about NSF's achievements." It's also good to generate press releases and send in white papers. Staff scientists at funding agencies often find themselves expected to be fluent in research areas far from their own original personal research interests. White papers and press releases can teach them the background they need to understand the new field they have suddenly been cast into. Press releases also provide evidence that their projects are successful, evidence that politicians and political appointees—the people they need to impress—can understand.

There are other ways to make your funding-agency staff happy. These staff members often spend a fair bit of time organizing proposal selection committees—a thankless task that generates nonstop complaints. So when they invite you to serve on committees, you might want to tell them you'd be delighted to help and that you appreciate the effort they are putting into the proposal-selection process.

CONSUMER: Press officers

MAIN CONCERN: "How can I keep my institution in the news?"
I have had many enjoyable conversations with Robert Naeye, editor-in-chief of *Sky and Telescope* Magazine, about science communication. Naeye explained to me that science writers mostly come in two varieties: there are journalists who work for magazines and newspapers and so on, and there press officers and science writers who work for universities and government agencies. For example, your institution probably employs a press office or science writer whose job is to promote your institution.

The press officer at your institution represents a whole chain of consumers. Traditionally, she writes a press release and markets it to journalists. The journalists market it to their editors (or producers, in the case of TV). Then the editors, or the journalists themselves, present it to the public.

But this chain and this process are changing. Often these days, the process consists of a journalist cutting and pasting material from your press release. And sometimes the "journalist" is just some robotic software that searches for press releases according to keyword, and then cuts and pastes them into various websites. Press officers also send their writing directly to websites like Digg.com and to bloggers, who can be quite powerful. Some run their own blogs that need feeding once a week or even once a day. Others are getting more involved in video, as this element of journalism becomes ever more important.

I find that science writers generally have broad scientific knowledge—and often surprisingly deep scientific knowledge as well. They love to share exciting science news. They generally appreciate what you do, and they may delight in telling you something you don't know about science or science communication. I suggest you let them.

CONSUMER: Reporters

MAIN CONCERN: Finding an appealing story about something unexpected—ideally a story with a protagonist and an antagonist

Besides the communications experts at our home institutions, we also interact with reporters who are not strictly on our home team; let's talk about those now. Almost every day, a reporter has to find a new story. There's a reason people use the word "story" to describe news: it's because simply describing the facts alone does not suffice. To grab people's attention and prevent them from changing the channel, a reporter has to spin some kind of tale, or at least provide readers with the props and encouragement to spin tales themselves. The best news stories contain something unexpected or ironic: man bites dog, as they say.

Reporters work on tight schedules. If they need you, they need you right away. So if you want to work with them, you'll have to make yourself easy to get hold of, and you'll have to return their phone calls immediately.

Now, reporters are interested in the truth. And science reporters, in particular, are interested in science. The trouble is that in the race to assemble a story quickly, particularly about a difficult and contentious subject like science, it's often easier to call something a controversy and present both sides than it is to figure out the right side to present. As Cornelia Dean, former science editor of the *New York Times*, put it, "Journalists will be hard-pressed to tell the good idea from the specious. And when we journalists are unable or unwilling to judge for ourselves, we often fall back on he-said-she-said reporting."[2]

When people talk about "manipulating the press," sometimes what they mean is that someone is delivering to the press what it craves: a controversy. For example, they find an event that will attract some media attention, and they generate controversy by opposing it. That turns the event into a more complete, easy-to-tell story, with a protagonist and an antagonist.

Here's a true story that illustrates the problem:

Jim Kukral woke up one Tuesday morning and decided he wanted to be on television. Somehow, anyhow. It didn't matter what he was doing on TV, what he was saying, or how he looked. He just wanted to be on television.

Kukral was not a scientist, a celebrity, or someone likely to have any particular influence over the media. But he gave the matter some thought. As he explained on his blog, "I asked myself, 'What does a reporter want right now?'" Kukral figured it out, put it together, and handed it to the local television stations on a silver platter.[3]

Kukral is from Cleveland, home of the Cleveland Browns, a football team with a terrible record—years of losing seasons—but a cadre of diehard local fans. The Browns' last home game of the year was coming up on Christmas Eve. There was sure to be some interest from the press about this game.

For the cost of ten dollars, Kukral registered a domain name called "Fanprotest.com" using the Godaddy webhosting service (www.godaddy.com). He threw together a one-page website. Then he put on his Cleveland Browns jersey, turned on his webcam, and recorded a two-minute video of himself, sitting in his living room, halogen lamp and couch in the background. In the video, he spoke passionately and seriously. He described himself as a lifelong Cleveland Browns fan, and called upon fellow fans to boycott the Christmas Eve Browns game in order to "make a clear point to the ownership of this team that we no longer are going to tolerate a losing team."

Kukral uploaded the boycott video, for free, to Google and to Youtube, then looked up all the local TV news stations in Cleveland and found their contact details, and sent each one an e-mail with a short summary of what he did and a link to the video. That was his entire effort. He estimated that it took him about an hour and a half.

Kukral's blog explains that he sent the e-mails out at around 3:30 p.m. and then went back to his normal workday. "I was chatting with my wife and playing with my kids when the phone rings, my wife picks up, she looks confused. She says, 'Scott from Channel 3 news is calling and wants to talk to you about some fan-protest story?'" An hour later, he had a television crew in his kitchen filming him. The story appeared that night on the 11 o'clock news in Cleveland, on WKYC.

Kukral used the news story of the Cleveland Browns' Christmas game to get himself attention. He could equally well have chosen a news story about climate change or ecology or evolution or stem cells to attach himself to. And the world is full of attention-seekers with fewer scruples than Kukral (who meant no harm); to some degree, whenever there is a serious point to be made in the news, there is an attention seeker ready to jump into the spotlight alongside it, in the manner of Kukral's prank. So if you want to make a consequential scientific point in the media, you have to prepare yourself by thinking ahead to what the attention-seekers might do to fill the niche you are creating.

Or perhaps you will take Kukral's place in this scneario—you will find a way to attach your scientific point to a news event, reacting quickly and visibly and presenting an opposing point of view. You can get a lot of attention by signing up to play the missing antagonist or protagonist on the world stage. If you do opt to use this powerful approach, I beg you as a colleague: do it responsibly.

CONSUMER: The public

MAIN CONCERN: To *feel* something

A major issue in ecology right now is how ecosystems will react to climate change. Many conservation scientists are advocating that we should design and maintain landscape-scale habitat corridors to support healthy ecosystems that will be more resilient to climate stresses. The idea is to help wildlife evolve or adapt by allowing, for example, animals and plants to migrate in response to changing conditions. During his election campaign, President Obama responded to this issue by, among other things, promising to "help animals adapt to climate change."

Steve Inskeep, host on NPR's *Morning Edition*, recently revisited President Obama's promise; he did an interview with political pundit Bill Adair about which campaign promises Mr. Obama had actually kept. When the campaign promise about maintaining habitats came up, Inskeep laughed. He asked Adair if President Obama had misspoken about helping animals adapt. Adair said, "We decided that that meant air-conditioners for bears."[4]

I have to admit that when I listened to the show and I heard President Obama's promise to "help animals adapt to climate change," my immediate reaction was similar to Inskeep's misconception. Before I had time to think about ecology, a certain image of "animal" popped into my head: the cage-bound giraffes and bears and monkeys you see in a zoo. It was several seconds before I was able to move beyond this childish notion to an image that fit better with the ecologists' point of view, like a herd of pronghorns trying to migrate across a fenced and changing landscape. Eventually I caught on to the serious intent of the phrase—and to the creative thinking that inspired the strategy—but not before some lobe of my brain had registered a tickle of humor.

I think that we scientists often imagine that we can communicate with the public simply by reasoning with them. But then we find our attempts blocked by misunderstandings like the one above. As my childlike response showed, when ecologists talk about animals adapting, they have a different reference frame than non-ecologists. An even thicker language barrier than that often exits between scientists and non-scientists. A serious plea can easily provoke giggles, and turn the audience we are trying to reach into a room of restless teenagers trying to avoid an algebra lesson.

So I want to emphasize here a particular public desire that we can keep in mind to help us past these communication hurdles. Not everyone wants to delve into a reasoned argument, and not everyone wants to take time to learn the languages and frameworks of our particular scientific subfields. But everyone wants to *feel something*.

We like to feel joy, pain, guilt, fear—any emotion, really. That's why we go to the movies and listen to country songs. That's even partly why we pay attention to news stories. To grab hold of the public's attention, we can use humor, fear, love, or sex appeal—we can use drama, stories, sounds, or pictures. But I think the key is that the first connection should be emotional or instinctive, not rational or cerebral. Poet and memoirist Maya Angelou summed it up: people will forget what you said, people will forget what you did, but people will never forget how you made them feel.[5]

CONSUMER: Policymakers

MAIN CONCERN: Making decisions quickly, and moving on

In *Escaping the Ivory Tower*, Nancy Baron paints a picture of the life of a United States senator. What shocked me about this picture is how little time such an important decision maker has to learn about any given issue that he or she has to vote on. For example, each year more than 10,000 bills are introduced in the U.S. Congress. A legislator may vote on more than 500 of these bills—in less than 150 working days. While our research time as scientists is mostly spent looking into all the possible ramifications, details, uncertainties, and caveats of our experiments, a senator's time is really spent mostly on meeting with people and making decisions based on what he or she learns in those meetings. There is no time to look into the caveats. Even the senator's staffers can barely keep up.

So to talk to a policymaker or one of his or her staffers, it's crucial, foremost, to be brief. You'll have to skip the background material and qualifiers that your scientific colleagues might expect from you. As Barron puts it, "Plan on five minutes, hope for fifteen, and dream of thirty."[6]

Also, everyone who walks into such a meeting has what's called an "ask," some kind of request to make of the policymaker. For scientists, the "ask" is usually for more research funding of some kind. You can be sure that this request for research funding has been unenthusiastically anticipated; if you do make it, you will probably elicit only a sigh or a furrowed brow. A better approach is not to ask, but instead to deliver a short summary of an issue that's relevant to the policymaker, a concise opinion on what should be done, and an offer of your services as an advisor, should more input be needed. The magic words here are "the bottom line is . . ." or "our educated guess is. . . ." We'll talk more about this topic later.

Focus Groups: A Way to Understand Your Audience's Needs

When I pitch a song for a country artist, I want to hear someone tell me that the artist loves the song and will record it right away. Barring that, I'd like to hear that the artist loves most of the song, and if I just tweak the lyrics to the bridge, it will be perfect for her next album. However, all I generally get from the artist's manager is "pass" or "hold." A hold means "We might stick it on the pile of fifty or so songs that we're going to listen to a second time." A pass means "No thanks." If nobody returns your calls or your e-mails, that's also a pass.

While you are in school, you generally get lots of feedback, in the form of comments and grades. But in the music business—as in most businesses—by and large you don't get feedback. You get nada. If you're at the stage of your scientific career where you are applying for a lot of jobs or grants, you know that this situation prevails in the business of science too. It can be hard to figure out what your intended audience wants, even if you have done your best to make a list of their needs, like the list we made above.

One technique marketers use to solve this problem is to test their ideas on a focus group. A focus group is a group of people representing your target demographic who will take the time to give you feedback on your products. When you are about to give a talk somewhere, you might practice in front of scientists at your home institution and watch how they react. Or when you are writing a proposal, you may run your ideas by your proposal team to get feedback. Sometimes scientists will even assemble a "red team" review, where a panel of local experts reads and troubleshoots a proposal. The term "red team" comes from the military, where it means a group of friendly soldiers who stage a mock attack on a base to test its defenses during a war game.

Sometimes we scientists find this process to be as painful as an actual war, not a war game. Or sometimes we take the attitude that we should control the focus group and run the show. We wilt under criticism, remembering the pain of seeing red ink on our homework papers.

When I'm in this situation, it helps me to keep in mind that besides helping us hone our ideas, focus groups can be vital for spreading them. Calling together a focus group means asking the members to do you a favor. And according to the Benjamin Franklin effect (chapter 4), when people do you a favor, it makes them form a higher opinion of you. The members of the group you're testing your product on might even recommend your product to others. That was how TCHO, the chocolate company we talked about in chapter 4, started out.

Focus groups are yet another example of a marketing technique already used regularly by scientists. We will see still more examples of these in the next few chapters.

CHAPTER EIGHT

Our Products: How We Get Job Offers and Funding

Before the Internet revolution, people used to talk about the "four Ps" of marketing: product, price, place, and promotion.[1] Now that customer relationships and even customer participation have become such an important component of marketing, some business book authors have stretched the four-Ps gimmick, adding Ps or other letters to try to cover the new ground. But one thing hasn't changed. A crucial step in marketing is still the first of the four Ps: figuring out what the products themselves should be.

We talked earlier about how engineers design automobile doors to produce the right solid-sounding *thunk* when you close them (chapter 3). We talked about TCHO, the chocolate company that sends feedback from its customers to its growers (chapter 4). We talked about Seth Godin's concept of the purple cow: a person so passionate, or a product that is so remarkable, that people can't help talking about it (chapter 5). These examples show how designing products is an intrinsic part of marketing.

So let me ask you a question that I've asked myself many times: as a scientist, what are your products? To summon the full power of marketing, we scientists ought to design our products with the consumer in mind. But it's not obvious to begin with what our key products are.

As a graduate student, I was educated to think that my most important products were journal papers. Papers certainly are the bread and butter of a life in science or academia, and a long list of publications on your CV helps impress a hiring or tenure committee. But I think that a management consultant examining the careers of scientists would not see papers as our primary products. That's because nobody pays us for our papers, at least not directly.

The same kind of question comes up in the music business: what is the product? It used to be that albums were the chief product of the music industry. Artists made a bit of money playing concerts on tour, but the real money came from selling millions of fifteen-dollar CDs. The songs played on the radio and by the artists on stage were marketing tools that advertised the CDs.

Then along came the MP3 and Napster and iTunes, and many people stopped buying CDs. There was no longer a need to buy a whole CD full of songs you didn't know, when you could just buy the songs you liked one at a time—or easily (though illegally) copy them from friends. This change left record labels scrambling to find new ways to make money.

Now, hundreds of lawsuits later, most of the music industry has adjusted to the idea that its primary moneymakers are concert tickets and merchandise sold at concerts.[2] CDs and MP3s—as well as the songs themselves—have essentially become marketing tools to help sell other products. This concept has been hard for many musicians to swallow. But it was necessary for the music industry to take a fresh look at the situation and come to grips with it, so they would stop hitting their heads against a wall, initiating lawsuits that alienated consumers but didn't affect how consumers acquired songs. It seems to me that scientists might benefit from a similar introspective look.

Two respected marketing pundits actually took time away from the lucrative business world to consider the question of scientists and their products. J. Paul Peter and Jerry C. Olson, authors of the textbook *Consumer Behavior & Marketing Strategy*, wrote an article for the *Journal of Marketing* called "Is Science Marketing?"[3] In this article, they argued that the main products of science and scientists are theories; they describe a scientific marketplace where theories carry a kind of price tag, just like a lamp at a garage sale. Low-priced theories, say Peter and Olson, emerge effortlessly from the existing worldview and the research skills of the target scientists, without demanding an intellectual stretch. Higher-priced theories are those that would require denouncing one's own papers or retooling the lab. In Peter and Olson's view, journal articles and conference proceedings and talks and so on are marketing tools to promote theories at all price points.

Part of Peter and Olson's concept rings true to me. Papers, talks, and conference proceedings do feel to me like marketing tools—but they're not the products themselves. Your product should be something that many people would be likely to pay for. I don't think you and I could get people to pay us much to read our conference proceedings or attend our technical talks.

Where I think Peter and Olson made their first mistake is using the word "theories." A theory is an established framework that encompasses a large body of evidence. I think Peter and Olson meant "hypotheses," which, as you know, are more like challenges, or implied questions to be answered through research. There is a steady need for scientific questions and tools to answer them with; we'll talk about that in a bit.

But more importantly, I want to point out that neither theories nor hypotheses nor research tools directly bring money to scientists. Comparing the science business to the music business, I'd have to say that none of these creations is exactly analogous to the tickets and T-shirts that musicians sell on their concert tours, which actually bring in the cash nowadays. I claim that there is one product in particular that makes the scientific world go round: proposals. Some of us make some money writing books and articles, or cashing in on patents and inventions. But proposals represent the bulk of the livelihoods of scientists at all career stages.

Many scientists are supported solely by proposals they write. Proposals fund the work of graduate students, postdoctoral fellows, engineers, and other support staff as well. Even if you spend most of your time teaching, you are still buoyed by proposals, the proposals that launched and continue to maintain your department; these may be internal or philanthropic or political in nature, but they are still proposals. Ultimately, everyone in the world of science needs to think about proposals, whether they are writing them themselves or not.

Let's take the idea that proposals are the staple product of the scientific economy and see where it leads. I think you'll find that this viewpoint lends a kind of satisfying order to the scientific marketplace.

The Proposal Supply Chain

The vast market for automobiles drives markets for car parts (tires, windshields, air bags) and raw materials (rubber, steel, plastic, and glass). The network of producers, distributors, warehouses, and vendors that supply

The Proposal Supply Chain

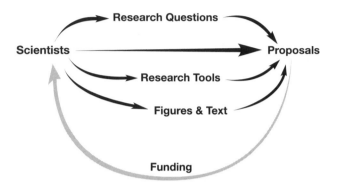

Figure 8-1   The Proposal Supply Chain. Source: Pete Yezukevich.

these parts and materials—together with the car manufacturers—is called the automotive supply chain. Scientific proposals also need a supply chain of materials that are components in their manufacture. One could view many of the institutions of science and academia as means to harvest, warehouse, develop, and ship these materials to the scientists who write the proposals.

A proposal requires a team of people who will do the proposed work. It needs a research question and a research tool that can answer this question. And of course, it requires some figures and text to communicate all of this to the reader. Figure 8-1 illustrates how these products and suppliers fit together in the proposal supply chain.

Of course, to actually get a proposal funded, we need to market our work. For this task, we employ marketing tools such as talks, papers, press releases, conferences, mailing lists, and websites. Some of the items on this list are what junior scientists might ordinarily be trained to think of as our primary products. But to me, these all look like either tools for marketing proposals, or links on the proposal supply chain.

Informal Proposals

When I say proposals are our primary products, I don't necessarily have in mind some kind of lengthy, official-looking document submitted in response to a proposal call. Sometimes a proposal can take the form of a white paper or even just a conversation (preferably one involving a prop). For example, I found online the slides from a PowerPoint talk by a staff member at a large government agency that has funded many science projects in the $4–135M range. The talk shows two mysterious bullet points about the funding opportunities at the agency:

- No budget line or proposal opportunities at approximately the $4–135M level
- Proposals in this range treated one at a time, overseen by program officers on an ad hoc basis

Let me translate this language. If you want a grant for less than $4 million, you can wait for a call for proposals, prepare a proposal that meets the specifications in the proposal call, and compete with your peers to convince an overworked review panel that you deserve funding. But if you want a grant for more than $4 million, all you need to do is go directly to your program officer and ask for it.

Of course, it's not that simple. You can't just wander in off the street and credibly ask for $50 million. But if you develop a real, honest long-term relationship with the people at the funding agency and show them that your idea meets their needs, you might find yourself writing a proposal in response to an opportunity that never existed before. It's crucial that the relationship be real and honest, as we discussed in chapter 4.

The situation is surprisingly common. My sources say that at some agencies, such as the Department of Defense (DOD), the Defense Advanced Research Projects Agency (DARPA), and many private foundations, program managers make most of the funding decisions directly. Even at agencies like the National Science Foundation (NSF) and the National Institutes of Health (NIH), where most funding decisions are primarily based on peer review, the program manager can have a surprising amount of power. At most agencies, if the program manager feels that the work meets the current programmatic needs, she can make the decision to override the rankings delivered by the review panel.

Now let us talk some more about the ingredients of proposals and the marketing of proposals, keeping the proposal supply chain in mind. As I mentioned earlier, sometimes the components of a proposal are marketing tools themselves.

Free Samples

I love to shop at Whole Foods. The knowledge that I am buying organic, often locally sourced foods makes me feel good about myself. I also like that the company pays its workers a living wage and offers them excellent benefits and health care.

But what really excites me when I walk in the door of my local Whole Foods is the thought of the free samples. I know exactly where in the store they place them: the red grapes, the all-natural chocolate sandwich cookies, and the Pirate's Booty snacks. My mouth waters and my tail wags.

Giving away free samples is the oldest marketing trick in the book. It simultaneously generates goodwill and educates the customer about the product. The marketing experts at Whole Foods know that happy consumers who have been snacking on grapes will want to go to the store more often, and that they will spend more time there when they do. And that's worth the price of a few grapes.

Here's a story from the music business about free samples. I like this example because, in this case, the free sample is an intellectual product. In 2007, when the music industry was recoiling from the shock of album revenues lost to the digital revolution, the U.K. band Radiohead did something that set the whole industry on edge. They decided to release their new album *In Rainbows* by making the whole album available for download online—for whatever price listeners felt like paying. The fans could choose to pay as little as 45 pence (about 1 U.S. dollar), just enough to cover the credit-card handling fees.

The members of Radiohead were among the first to realize that they stood to gain more in goodwill and future T-shirt sales than they would lose by undercharging for the album downloads; they understood that the recorded music was a marketing tool, not the product. Indeed, the (almost) free sample of the band's work turned out to be a spectacular marketing tool. First, the band presold 1.2 million downloads by the day of release. Later, when the band released a physical CD, it charted at number one in the U.S. and in the U.K., selling more than 3 million copies and winning a Grammy award for Best Alternative Music Album. I don't know how many concerts tickets they sold that year, but I bet it was a lot; Prince, Trent Reznor, Liz Phair, and Beck soon began experimenting with a similar approach.

To use this marketing technique as an academic, your instinct might be to give copies of your papers away. But marketing tools, like papers, like songs, are presumed to be free, so that doesn't score you very many points. To win real favor from our customers, we have to give away free samples of our actual products: proposals, research tools, and so on.

White Papers

What I mean by giving away a free sample of a proposal is writing a little white paper and showing it to your funding agency. A white paper is a kind of mini-proposal—snack-sized. Deborah Leckband, the Reid T. Milner Professor of Chemical and Biomolecular Engineering at the University of Illinois at Urbana-Champaign, explained to me how she uses white papers as free samples to market her proposals:

> You have to go down to D.C. and explain to them why this is the most fantastic stuff, and then if they are persuaded they will invite you to submit a white paper. I think I spent about three years communicating with my program officer at the Office of Naval Research before I got funding. I submitted a couple of white papers and they said no. Then, after the third white paper, they invited a full proposal, and they funded it. In the process, I got to know what they wanted and how to slant my proposal to fit what they needed.

There are no official rules about what a white paper should consist of. But here are some guidelines that I picked up in my interviews. A white paper could be one page or up to four pages. It should be divided into sections with section headings and no more than roughly a page and a half per section. Each section should have a few paragraphs of text, a figure with a caption and a few bullet points with sound bites about the topic at hand (we'll talk about sound bites in chapter 13). The text should generally be written at the level of someone who has a PhD but is not in your field of science. There should be plenty of white space—this is not an exercise in cramming words into a document with a page limit.

Unfortunately, writing white papers is another one of those crucial science-marketing habits that's seldom taught in graduate school. Maybe the closest most graduate students come to learning about this tool is when they write their applications for postdoctoral or faculty positions.

The three-page essays usually requested in these applications look to me more or less like white papers in disguise.

It's hard to overstate the importance of white papers as a science-marketing tool. David Seiler from NIST insisted, "You should always carry around a white paper in your back pocket." But the demands of relationship building must come into play here, too; you can't just hand out your white paper indiscriminately to everyone you meet. It seems to me that we must always get permission before making advances (see chapter 4), that is, we must wait for people to ask us about our work before we start handing them things.

A marketing tool that's closely related to the white paper is the "one-pager."[4] When you meet with a senator, congressperson, or congressional staffer, you're expected to leave a one-page write-up of your position on the issue at hand. You can think of this document as a kind of one-page white paper.

In any case, the most reliable route to a successful proposal (assuming you have a good idea) is to market your work *before* the announcement of opportunity even comes out. Stanford physics professor (and my life-long friend) David Goldhaber gave me some advice on this topic. He said, "I tell new faculty in our department to write an e-mail to the relevant program managers in all the relevant funding agencies. In the letter, introduce yourself, and ask what he or she is looking for these days." Then, see if you can put together a white paper about some research that meets your program manager's needs.

It is possible to be so admired by your funding agency that you find out in advance when a call for proposals is coming. Some scientists are even asked to help write proposal calls. If you do help write the call, you won't be able to propose, of course, but you can bet that the agency will look favorably on your work. As economist and author Harold E. Marshall put it in his interview with me, "Figure out how to be indispensable to everybody else and you'll always get money."

We will talk in detail about preparing proposals themselves in the next chapter. Now let's talk about the ingredients of a proposal and how to market them up the supply chain. Some of these ingredients can also be marketing tools when they are used as free samples.

Proposal Ingredient: Research Tools

A good proposal is about what happens when a research question meets a research tool; every proposal needs a research tool. Some examples of research tools are pieces of software, instruments and observatories, laboratory techniques, catalogs, and even equations. You probably spent most of grad school creating research tools or becoming proficient in their use. Ultimately, my business perspective would say that you undertook this training because you and other scientists need these tools to create winning proposals.

Research tools make great free samples. Research tools clearly take some care to craft, yet are often not themselves worth any money. So to some extent, we can feel free to give them away to potential customers, as long as we keep our names and logos stamped on them. This practice generates goodwill; it shows that you value the science and the community. It also educates your consumers about the nature of your real products—yourself and your proposals.

For example, if it's a piece of software or some numerical tool, you can give away last year's version of it for free on your website. (You might want to keep the latest version with all the bells and whistles for yourself.) Or if you are writing a theoretical paper, you will probably want to provide an equation or two that everyone can use without really reading the rest of the paper. Even equations can be made user-friendly for this purpose by putting them in convenient units and explicitly tying them to experimental or observational results. (See the section on elegance in chapter 5.)

Proposal Ingredient: Research Questions

Every proposal needs a research question. Sometimes a new experiment or observation poses a new question for theory to answer: "We noticed that some strains were much more drug resistant than predicted. What gives these strains their high resistance to these antibiotics?" Sometimes, the question comes from theory—usually in the form of a hypothesis or a prediction—for future experiments to address. "This theory predicts that more metal-rich stars will host more planets. Future observations of metal-rich stars will test this notion." Crucially, research questions sometimes go with buzzwords that encapsulate them, like "tachyon," "nanobacteria," "carbon planet," and "last universal ancestor"—all hypothetical notions.

There is a great market for provocative research questions, especially those with buzzwords. It came as a shock to me when I realized this, but in my experience, creating a new research question almost always generates more appreciation from your scientific peers than answering a question. An answered question leaves the realm of science, in a way, and comes under the care of textbook writers, engineers, and science teachers. If you really firmly answer a question, you almost threaten to put your peers out of business. In contrast, every proposal needs a research question. So if you can come up with a new question that needs answering, your fellow scientists will be grateful.

This is not to say that we don't aim to answer questions. But maybe this is why my advisors in graduate school insisted that I list some remaining unanswered questions at the end of every paper I wrote. In the business of science, launching new questions is just as important as answering old ones.

Proposal Ingredient: Students and Postdocs

I argued above that people are one of the key links in the proposal supply chain; every proposal needs a team of researchers to support it. This situation has an interesting ramification. It means we can fulfill the desires of our scientific colleagues by providing them with more scientists.

For example, at every career stage I've found that it pays to keep a mental log of the talent in your field, and to keep an eye out for smart young researchers whom you can recommend to your colleagues when they ask for your advice. You can develop a reputation as a kind of talent broker, and find yourself invited to participate in proposals or other activities as a result.

Proposal Ingredient: Ourselves

We ourselves are one of the links in the proposal supply chain; we scientists are products that we can supply to proposal writers to help generate money. Now, as a junior scientist, one of your main goals is to get a job. I think viewing yourself as a product, a brand, and a link in the proposal supply chain is one way to help understand how you can accomplish this goal. The idea is that you'll need to show the people whom you want to hire you just how useful you can be for developing and marketing proposals to bring in money.

One way to market ourselves is by giving away free samples. Giving away a free sample of yourself means working for free. That's part of what being an intern or graduate student or postdoc is about: working for free (or for cheap) to show off what you can do. I think I probably would have enjoyed graduate school more and done a better job at it if I had pictured it this way.

Another way to market yourself as a junior scientist is to show that you have experience writing proposals yourself. That directly demonstrates your potential for bringing in money, and it can rapidly expand your circle of collaborators and potential advocates. Often these days, postdocs will obtain some experience writing proposals, even winning and managing small grants.

But back to the analogy with the automobile business—you might say postdocs writing proposals is equivalent to a tire company deciding to sell cars. Tire companies might hesitate to sell cars because that would put them in competition with their key customers, the automobile companies. Likewise, in science, postdocs who decide to write proposals can find that they are suddenly in competition with their mentors. Some mentors and institutions encourage this competition; others do not. I myself was lucky to have mentors who supported and encouraged my proposal-writing efforts—though I have heard of one advisor who turned as red as a can of Coke when he found out his postdoc was assembling a proposal team.

Probably the best path to job-hunting success is to try to become what's called a "go-to" person in one small corner of science. If you can become the top name in a subfield—the person others have to go to for research tools, invited talks, review papers, and so on—then it goes without saying that you will be a key person for any kind of proposal, large or small, that involves your field. And if you are crucial for proposals, you must be a crucial person to hire. Becoming the top person in some subfield also helps the people who are writing you letters of recommendation. It allows them to speak in the language of the hero/warrior archetype, using words like "best" and "top." Even "one of the top two or three" is a good phrase to see in a letter of recommendation. This is the way I see most junior faculty get hired.

A go-to person is just another name for the brand leader, like Coke,

Hertz, Xerox, and so forth. When you decided where to go to school, your decision was probably based partly on the school's brand name. When you apply for a job, you may find those same universities are deciding whether they want to choose you, at least partly, on the basis of your brand name. It's only fair.

The Signature Research Idea (SRI)

In popular music, most artists first become known for one or two hit songs that define their style in the public eye. Gretchen Williams's song "Redneck Woman," which I mentioned in chapter 5, is one such example. She has written and sung many other songs. But she will always be known as "the Redneck Woman."

The same phenomenon occurs in science. We may steadily create research tools and research questions, but only a few of these research ideas become real hits. These few hits can fuel a string of proposals—even launch a whole career. I'm going to refer to this kind of hit research tool or research question as a Signature Research Idea (SRI).

If you have an SRI, it becomes, in a nutshell, your résumé. A hiring committee looking at your record doesn't have to work too hard to justify hiring you. All they need to do is say you're the one with a certain SRI. It's the equivalent of when your friend asks why you bought tickets to go to a Britney Spears concert, and you say, "Oh, she's the one who sings that song 'Womanizer.'"

Caltech physicist and author Sean Carroll captured this idea in his article "How to Get Tenure at a Major Research University." Carroll said, "It's easier for people to think about what you've done if it can be summed up in a sentence. When people ask, 'What was your major contribution?' have an answer ready." He went even further, suggesting that the best kind of SRI is one that directly bears your name. "A slight exaggeration, but if you have something named after you—a theorem, an experiment, a model—it's a big help."[5]

My first signature research idea was the "band-limited mask" I mentioned in chapter 5. I had lots of other ideas as a postdoc, and I wrote lots of other papers, but most of them got buried, appropriately, in the dustbin of science history. This one, however, became the heart of several proposals, and it still lives on. Because I had the SRI "band-limited mask" associated with my name, I became a go-to person in a small communi-ty—a subfield—of band-limited-mask users. That calling card opened the doors to the next stage of my career. Since then, I've attached my name to a few other SRIs, which you can easily find if you feel like Googling me. If I'm lucky, maybe one day I will top all my SRIs with a really big hit, something on the order of "PCR" (Polymerase Chain Reaction), "seismo-graph," "general relativity," or "evolution."

From Backup Singer to Rock Star

Let's look further at how careers in the music business work. Different products are important to musicians at different stages in their careers. At first, aspiring rock stars may have only themselves to offer; they can play in a backup band or sing other people's songs. Later on they may con-centrate their efforts on making and selling recordings, and try to score a hit song. But they still generally make little money at this stage, and they often go into debt. If they work hard and are fortunate, eventually a small number of them become big-enough brands that they can make real money selling the music business's biggest money-making product: concert tickets. That's when they've made the big time.

Likewise, different products are important to scientists at different ca-reer stages, as illustrated in Figure 8-2. If you are a student, your only product may be a warm body who can carry out the work in a proposal, i.e., yourself. You'll work as a backup singer in a band of scientists, de-veloping a brand for yourself as a hard worker, a fast learner, and so on.

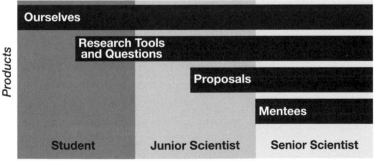

Figure 8-2   The products we create at different stages of our scientific careers.
Source: Pete Yezukevich.

Then you'll start making your own records, so to speak, trying to develop the hit songs (SRIs) demanded for use in proposals. These are our inventions, our discoveries—DNA, black holes, lasers, Higgs bosons, micelles, nanotubes, and so on. One day, maybe you'll make it big, and become a senior scientist. Then you'll bring in the cash to pay your own band (research group) by writing proposals. At this point, you may also work to promote younger people—such as students or postdocs.

How do you become a brand leader, or a go-to person? By inventing or discovering and then promoting SRIs. In other words, to become a rock star, you need a hit song.

Promote the Idea

Earlier on, I made the claim that the best marketers are simply the most successful scientists, not the ones you think of as marketing themselves. That's because a good marketer doesn't exactly promote herself—at least not directly. A good marketer promotes her products. If people come to believe in your products, they will come to believe in you too.

In science, it's often taboo to promote either yourself or your proposals directly, except through certain channels. But it's always good to promote our SRIs or our students. This kind of promotion does not generally appear to be self-serving—provided that these products really are useful or entertaining to other people. It looks more like a kind of teamwork; you and your SRI or you and your student become a kind of a team, supporting one another.

One of the scientists I interviewed about marketing was David J. Pinsky, chief of cardiovascular medicine and scientific director at the University of Michigan Cardiovascular Center. He had a kind of philosophical take on this notion. "It's really about marketing the idea. Some people get confused between the idea and themselves. Why things happen isn't because we make it so, it's because it is what it is. So don't market yourself, market the idea."

It is what it is. A beautiful sentiment coming from someone who has looked, literally, into many people's hearts.

CHAPTER NINE

Writing Proposals and Making Figures

 I received a fair bit of advice from my colleagues in interviews and on Facebook about how to write a good proposal. That's what I want to share with you now. Also, proposals seem to demand certain kinds of plots and figures to make their points; plots and figures are sort of like raw materials in the proposal supply chain. So we'll talk about making figures in this chapter as well. There are many books about proposal writing, and one can take courses on it. But I'm going to try to couch this discussion in terms of the marketing and business language we've been discussing, and continue to see how far I can push my business analogies.

Your Funding Agency and the Review Panel: What's In It For Them?

I heard over and over from my colleagues that marketing a proposal means thinking about the needs of the funding agency that put out the call and the panelists who are reviewing the proposals. That certainly fits to a tee the definition of *marketing* we talked about in chapter 2. As Anne Kinney, who used to be the director of the Astronomy and Physics Division in the Office of Space Science at NASA Headquarters told me, "You need to make it easy for those guys to fund you, and you need to do that by showing them how your specific science scratches their itch."

Moreover, your poor colleagues who are sitting on the proposal review panels also deserve your attention. Panel members are usually feeling pressed for time and burdened by the large stack of proposals they need to read. They may feel disheartened because of the many proposals they will have to reject. They need you to help make the process efficient and fun—or at least as pain-free as possible.

For example, I heard many times that you shouldn't make the panel fish around in the proposal trying to figure out what you plan to do; you have to describe your main goal at the very beginning of the proposal. In other words, you should probably say literally, in the text, "We propose to . . . X, Y, and Z." If you feel you can't write the words "We propose" because you're the only one on the proposal team, you can write, "I propose" instead. In practice, the wording "I propose" usually only comes up on job and fellowship applications.

Proposal writers often feel like they need to push lots of words and ideas at the review panel. They write as many words as possible to bolster their case. They describe as many applications for their tools as they can. They claim they will solve every relevant problem under the sun.

But that approach doesn't make life simple for the review panelists. It makes discussing the proposal complicated and tedious, and hampers the panelists when they try to explain their decisions to the funding agency. It leaves the panelists with no opportunity to show off their own expertise and creativity by telling their own stories about the many possible applications of your proposed work.

I have been told that the best proposals take a different approach. They state one single research question and match it with one single research tool that will answer the question. Malcolm Fridlund, at the European Space Agency, wrote very clearly about the need to focus your proposals in this fashion. Fridlund is the scientific manager of the Darwin project, a proposed multi-billion-dollar international space mission. He said it's probably best to ". . . focus your proposal around one simple, testable main idea. Then ask for funding for testing it in a yes/no mode." That makes it easy for the review panelists to explain the proposal to each other and to the program manager, and for the program manager to explain to his manager as well. Fridlund's own favorite project, by the way, is about testing whether there are other planets out there that resemble the Earth. Other Earths: Yes/No?

Given that your proposal is going to be about one single question, it seems important that your proposal make it clear that your work will really answer the question at hand. I don't mean address the question or make progress toward answering the question or be relevant to the question—I mean *answer* the question. For example, let's say you are proposing to use a new instrument to study a special kind of cell called a "wangocyte." Your question is "Do wangocytes have mitochondria"? You should probably propose to examine enough wangocytes to answer that question with acceptable statistical confidence—no more, no less.

Maybe you or other wangocyte researchers will also be able to use your new data to do some other study, like make a contribution to our understanding of vascular disease. But if you describe this second study in your proposal, you probably won't have space to explain and justify it in detail, and you'll end up with two stitched-together bad proposals instead of one good proposal. You can easily confuse the poor overwhelmed review panelists and dilute your message. It's better not to mention one of these studies, and just let the review panel members fantasize about using your data for other purposes. That way, the conversation in the review panel room will be about what else your proposed work is good for, not about whether or not you can actually answer all the various questions you posed.

Scientists have advised me that proposal writers need to focus their words too, not just your ideas. Nicholas Suntzeff was the director of the astronomy program at Texas A&M University and vice president of the American Astronomical Society; he now works at the U.S. Department of State. Suntzeff told me, "Most proposal writers feel they have to max out the scientific justification. This is not necessary. Instead, use white space, tables, and figures to allow the reviewer's eyes to rest."

Sometimes proposal writers use alliteration or rhyme or some other gimmick to catch the eye of the reader, especially in the title of the proposal. That trick, I am told, almost always backfires, because it's such an obvious ploy. Many people told me: don't be cute or silly when you are asking for thousands of dollars.

But Suntzeff also said, "Put something memorable in your proposal—a figure, a comment." The idea is that if you can think of a way to say something legitimately clever or entertaining in the proposal, it might help your chances, because it will break the tension in the review panel. A wry, self-deprecating comment in the body of the proposal might have the review panel eagerly passing your proposal around and chuckling.

The Three Kinds of Figures Every
Proposal—and Every Scientist—Needs

Besides the title and the abstract, the most important parts of any proposal are probably the figures and the figure captions. That's because when readers are in a hurry to skim through a proposal, they naturally start by looking at the figures. So, like research questions, research tools, and personnel, figures are a crucial commodity in the scientific marketplace. Of course, besides proposals, every talk and every paper needs lots of figures as well. So there is a steady demand for them from scientists at all levels, and science writers too. I have found that creating a "hit" figure—one that becomes very popular—can bring you just about as much glory as a well-cited paper.

Sometimes, a hit figure is not one that would meet the standards of a scientific publication. For example, I once took a figure from a classic paper by Harold Zapolsky and Edwin Salpeter, and spiced it up by adding color clip-art images of Neptune, Saturn, and Jupiter to show how those planets conformed to the models shown in the figure. Other scientists started borrowing it to use in their talks. Soon other modelers were improving upon it and updating it, but for a while, it was known as the "Kuchner" diagram, despite the fact that I didn't put my name on it (I credited Zapolsky and Salpeter, of course). Thankfully, Zapolsky and Salpeter didn't seem to mind.

I've come across three types of figures that seem to show up over and over again in talks and proposals. I'd like to suggest that these are the kinds of figures every scientist needs—the kinds that you probably want to think about whenever you sit down to make a new figure. Let me describe them to you now. For more ideas about how to make an effective figure, I recommend Edward Tufte's book, *The Visual Display of Quantitative Information*.

The Beautiful Butterfly Figure

To start with, many proposals start with one figure whose purpose is just to grab the reader's attention, a pretty picture like the spectacular images you see from the Hubble Space Telescope or the dramatic artists' conceptions and colorful models you find on the covers of *Nature* and *Science*. Glamorous pictures like these—I call them "Beautiful Butterfly

Figures"—also make good introductory slides for a talk. Several science writers have also told me that if you have a Beautiful Butterfly picture, you may have the makings of a good press release, even if the science itself is not world changing.

A Beautiful Butterfly picture doesn't have to carry any quantitative information at all. In fact it's probably better if it doesn't. There is a time and a place for people, even scientists, to react purely with their emotions. If people are squinting at a color bar or axis label, they will be unable to do that.

The Family Portrait Figure

Next, your proposal might need some kind of figure that sums up the contributions of everyone else in the field, a figure that plots as many people's data or theories on it as possible. For example, maybe you could use a log-log plot that spans 10 orders of magnitude and shows the work of five different research groups, like Figure 9-1. Either that or you could make a diagram of the object you're studying labeled to show all the processes going on in it—and all the names of the people in the field who are studying those processes. If you can make a figure like this, then all the people whose work you plotted will suddenly be interested in you; all five research groups will want to show this figure in their talks. I think of this

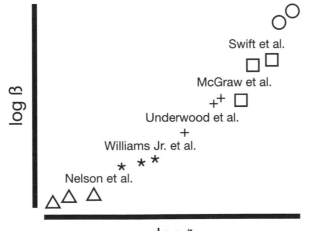

Figure 9-1   The Family Portrait figure shows a little bit of data or theory from everyone in the field and thus makes everyone happy. Source: Pete Yezukevich.

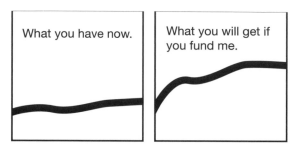

Figure 9-2   A Jenny Craig figure illustrates the impact of your work, like those ads showing people before a weight-loss program and then after the program. Source: Pete Yezukevich.

kind of figure as a kind of "Family Portrait" of everyone in the field, albeit one that shows everyone's data, not everyone's faces.

The Jenny Craig Figure

I think this third kind of figure, illustrated in Figure 9-2, is the crucial one for selling a proposal; absolutely every proposal and every white paper must have one. Mario Perez, a scientist at NASA headquarters, calls this kind of figure the "Jenny Craig figure," because it's just like the before-and-after photos you see in commercials for weight-loss programs (like Jenny Craig). To make a Jenny Craig figure, first show your readers an example of what they have now, and next to it, an example of what they can expect to get if they buy your product, i.e., fund your proposal. In other words, the Jenny Craig figure compares your results or your predicted results with the previous or current state of the art.

You can make this comparison in two panels, like Figure 9-2, or in one panel, with different colored lines or symbols. Just make sure there's a major difference between them—make sure your customer can easily tell that the participant is skinnier *after* the weight-loss program. When I am writing a proposal, I often start by first coming up with the Jenny Craig figure. Then I write the text to go along with the Jenny Craig figure. The Jenny Craig figure serves as a kind of shorthand for the SRI (Signature Research Idea—chapter 8) that's behind the proposal.

Four Common Kinds of Bad Proposals

There are many ways to write a selfish, unfundable proposal. I've sat on many review panels and seen many of them—maybe you have, too. Some

common kinds of bad proposals reappear again and again. I think it helps
to see some of these common blunders spelled out.

The Social Security Proposal: We did lots of work on this subject
in the past. Therefore you should fund us some more.

The Fishing Expedition: We love this particular project, although
it's technically impossible, or unlikely to yield meaningful results. Give us funding to do it anyway.

The Big Toy Hammer Proposal: We built a complicated model or
experimental tool (a big toy hammer). It can be applied to many
problems and scenarios, all of them poorly described. But fund
me anyway so I can run around and whack things with my big
toy hammer.

The Vague Proposal: We want some funding, but we can't tell
you exactly what we're going to accomplish with it. Pay me,
please. Please?

A good proposal contains a well-posed and interesting research question paired with a solid research tool that will clearly answer the research
question. It is supported by a qualified team of people and a Jenny Craig
figure. Each of the bad proposals described above fails to deliver one or
more of these key ingredients. The *Social Security* proposal focuses too
much on selling the team, and neglects the research tool and/or question. The *Fishing Expedition* focuses too much on the research question,
and neglects the research tool. The *Big Toy Hammer* proposal focuses too
much on the research tool and neglects the research question. The *Vague*
proposal is probably missing a Jenny Craig figure.

To Sell a Blender, Show a Picture of a Margarita

I have one final and only slightly redundant comment about how to write
a proposal. It's easy when you are working on a proposal to put all your
effort in describing what you've done in the past. But what reviewers and
program managers tend to care more about is the future, the likely outcome of your research.

It's like when you're shopping for a blender. Unless you happen to be
a blender engineer, you don't really care precisely how the blender itself
was made; what's on your mind is what you're going to make with it

when you get home. You're not fantasizing about the oily gears and coils inside the blender, or even the glorious history of the appliance company that made the blender. You're thinking about a big slurp of something wet and frothy. Appliance manufacturers understand this; they know that if they want to sell you a blender, they should show you a picture of a margarita on the box.

Including a Jenny Craig figure is one way to make sure your proposal illustrates the smooth, delicious science your program will yield; it illustrates how your work will redefine the state of the art. Besides including Jenny Craig figures, I also suggest paying careful attention to the content of the text and the figure captions in your proposal, and using them to spell out, front and center, how your research will change the world, and how your scientific community will benefit from the change. Make the benefits you are claiming simple to summarize and easy to swallow. And they will buy your blender.

CHAPTER TEN

Papers and Conferences

 Something happened in the music business during the last ten years that we don't often get to see happening in science: the entire industry turned upside down. The primary source of money in the music business changed from CDs sold in stores like Tower Records to concert tickets and other touring revenue. What was the number-one product—recorded music— became primarily a marketing tool to sell other products.

I mention this upheaval again now because this is a chapter about journal papers and conferences. Presently, the life of a junior scientist tends to revolve around these two things. Writing papers and attending conferences are probably the two primary ways junior scientists market themselves to senior scientists, the people who can offer them jobs.

But after watching the music business change before my eyes, I wonder how long this situation in science will hold. Will journal papers and science conferences always have the same exalted place they do now in our lives? Have you ever attended a meeting on Second Life or Skype? If you haven't yet, I bet you will soon. It's not hard to picture a day when fuel prices or environmental concerns make conference travel prohibitively expensive.

The way we use journals has already changed dramatically. How often have you skipped reading a journal paper, used Google to find the illustration you want, and just pulled it right off someone's home page? Nowadays that's standard procedure, but it's easy to forget that ten years ago, you couldn't do that. And let me be the one to forecast this change: one day soon, our printed journals will be replaced by online video journals.

Already, the site Benchfly.com collects videos made by scientists that illustrate their techniques. Add a system for refereeing and a dash of prestige, and that starts to sound a little bit like a journal. Cisco Systems, the gigantic company that owns Linksys networking products, predicts that 90 percent of internet traffic will be video by 2013.[1] Some respectable fraction of that video traffic is going to be us scientists—and I predict journals are going to have to get in on the action, or risk losing even more attention from their audiences.

Let's take a closer look at today's journal papers and conferences anyway, and try to apply our marketing intuition to give us a better understanding of these institutions. They may be antiquated, or at least ready for an update. But these workhorses will likely be the models for the next generation of science-marketing tools.

The Journal Article

You are probably already an expert on how to write papers. So I don't want to bore you with a treatise on it. But I have two comments about papers, based on the marketing tactics we're been talking about, that I think you might find interesting.

First, the marketing viewpoint is that an academic paper is like any other product or marketing tool. It becomes popular not precisely because it's clever or correct, but because it meets other people's needs. You might think of your paper as a way to offer your colleagues ingredients that they can use for giving talks, writing more papers, and ultimately, writing proposals. These ingredients can include research questions, research tools, Beautiful Butterfly figures, Family Portrait figures, and so on.

Second, the marketing perspective tells us something about the audience for our papers. As you probably anticipated, part of your audience is the greater scientific community. But I think that the modern, relationship-centered marketing perspective reminds us about an important audience for our papers that we don't usually think of as part of the audience: our coauthors.

Your coauthors are already halfway through your marketing funnel. Writing a paper with them cements your bond. You hope that your paper will become indispensable to the greater community of scholars, and a classic in the field. But even if it turns out to be relatively obscure, your paper will still be useful for your coauthors. They get another publication to add to their list and a few more figures with their names attached to them. Don't have any coauthors? Unless your paper is "On the Electro-dynamics of Moving Bodies" (Einstein's first paper on Special Relativity), you may be missing an important opportunity. Maybe this is why Jim Austin, editor of Science Careers for *Science* magazine, tells me that single-author papers are "increasingly rare."

There's plenty of advice in the literature about choosing your coauthors. Peter Fiske, author of *Put Your Science to Work*, says that the best collaborator for an early-career scientist is someone who (1) is established, and (2) works at a different type of institution from your own (e.g., a national lab as opposed to a university), and (3) offers some new methodology to pair with your problem, or a new problem to pair with your methodology.[2] He suggests that 30 percent of a young scientist's yearly research activity should be with people outside his or her advisor's research group.[3] And of course, writing papers with your advisor is a crucial way to cement your bond with him or her. In fact, I didn't realize it at the time, but, looking back, it appears that I constructed my science career as a series of collaborations with senior people—hopping from one institution to another, collecting influential coauthors. It's a strategy that worked, and one I would recommend.

Finally, I like to think of every paper I write as supporting some signature research idea (SRI)—either launching a new one, or helping develop and promote one I've already launched. I'm not sure that this is the only route to success. But I have the feeling that, as in the music business, we scientists become known for our hits, not our album filler. So we need to conceive of our work as centered around creating, recognizing, and promulgating the very best research ideas, rather than dividing our time equally among all our various interests.

A sometimes-overlooked way to add impact to your papers is to submit your Beautiful Butterfly figure to the journal as cover art. Jay Morris, postdoc at the Medical University of South Carolina, and creator of the logo on the National Postdoctoral Association T-shirts (chapter 5) told me he was surprised at how easy it was to get his figure on the cover of a journal. "If you look on their webpages, most of the journals will allow you to submit an artwork idea for the cover. They don't want to have to pay people for the cover designs!" If you succeed in making the cover, you'll have a nice new achievement to list on your CV. And if you don't, well, there always seems to be a home for a good Beautiful Butterfly figure. Morris told me, "There was one time where I submitted artwork for the cover of the journal and they didn't use it, but later my advisor submitted a review article to another journal and submitted that artwork—and they used it."

By the way, I took a poll on my Facebook group about favorite paper titles. I was struck by the long list of clever, funny titles people responded with. My favorite came from Heidi Hammel, Co-Director of Research at the Space Science Institute in Boulder, Colorado, named by *Discover* magazine as one of the 50 most important women in science.[4] "We submitted a paper to *Science* with a title that we never thought would be acceptable, but accept it they did, to our great surprise and delight: 'New Dust Belts of Uranus: One Ring, Two Ring, Red Ring, Blue Ring.' And the acceptance date was Dr. Seuss's birthday."[5]

But many of the scientists in the Facebook group disparaged this approach. And a 2008 study by Itay Sagi and Eldad Yechiam in the *Journal of Information Science* had a panel of judges rate the "amusement and pleasantness" of the paper titles in two respected psychology journals: Psychological Bulletin and Psychology Review.[6] They found that while the pleasantness rating was weakly associated with the number of citations, articles with highly amusing titles (2 standard deviations above average) received fewer citations. Clearly, scientific cultures vary from field to field, and the interest in amusing paper titles may vary from field to field as well. But, I suspect that unless you are as prominent a scientist as Heidi Hammel, it might be safer to stick with straightforward titles for your journal articles.

In Hammel's case, using a joke title may even be a means to disarm her audience, which might otherwise view such a prominent scientist as snooty. Opening a talk by telling a joke is a related ploy. We'll talk more about this approach in the next chapter when we talk about giving presentations.

Organizing a Conference

I suggested in the chapter about building relationships that holding a professional conference is like hosting a party. It's a way to reach out and connect with many people all at once, and draw them down through your marketing funnel. Let's think about that analogy some more.

Often, scientific meetings aim to address an important question such as "What should the next big particle accelerator look like?" Or "How should the U.S. government prepare for the next pandemic?" It's hard to get into a party mood when such weighty questions are on the table.

I once went to a meeting whose stated goal was to "build consensus," but arguments raged all week about whose technique was the best. Then the person who convened the meeting simply put up a slide at the end of the meeting with what he wanted the consensus to be—when it was clear that there was no such consensus at all! About half the room left angry. That was not a good party.

This meeting left me wondering about the right way to handle such a meeting and produce the intellectual product you need, while still generating good feelings and building relationships. So I spoke to a friend of mine at a policy think tank who regularly organizes the kind of focused meeting where the stakes are high and the goal is to produce a written report with some kind of clear conclusions. I learned that managing a weighty discussion at a scientific meeting is like playing pin-the-tail-on-the-donkey at a child's birthday party. It's the most fun if everybody gets to win.

Here's what I learned about how to host a focused meeting, Washington, D.C.–style:

1. Interview the presenters before the meeting to find out where they stand.
2. Based on what you learn in the interviews, write up a first draft of the report. Call it a "strawman" draft to assure everyone that it is meant to be criticized.
3. Distribute the draft to everyone before the meeting, and solicit feedback. Use direct quotes from the participants, if you can, to show that you understand the ideas are theirs.
4. At the meeting, show everyone you have listened to their opinions, by reflecting them back at them.
5. Then, at the meeting, fine-tune the draft, incorporating inputs from everyone present.

In other words, you do most of the work of deciding what the outcome will be beforehand. Then, when people arrive at the meeting, they can relax and enjoy the party, because they already feel they've been listened to. This process rescues you from the impending clash of wills by avoiding the need to make big, stressful decisions in real time. If you follow these steps, then, when the meeting is over, you can walk away with an actual consensus, a finished product, and a lot of happy children. I mean scientists.

Inevitably, the room will divide into camps about what approach is the best to take on a given problem—as in the disastrous astronomy meeting I described above. I asked my friend how to handle this situation, and here's what I learned. When this happens, never ask the crowd which approach is better; that would either start a riot or cause people to clam up. Instead, ask each camp, "Under what circumstances would your approach be preferable?" This approach allows the camps an opportunity to come to some sort of agreement, and permits both groups to feel that their idea is the best—under certain circumstances. Everybody gets to play, and everybody gets to win.

Attending a Conference

Attending a conference is like attending a party. Now, I must admit that attending a party is something I must admit I only recently became truly comfortable with. But I think I know enough to flesh out this analogy. The idea is that you'll be popular if you help other people have a rollicking intellectual good time. That means introducing yourself to the people around you when they look lonely, and introducing them to other people whom you think they should know. It means stirring debate when the room is too quiet, and helping make peace when people get too rowdy.

Let's take the analogy still further. At a conference, when the topic I'm working on comes up, it's my special time—even if I'm not the speaker. It's the time to sit in front of the room. It's time to ask a good, fun, challenging question to engage with the speaker and get people thinking.

But although I should be used to these moments by now, I still feel just a little queasy when my topic comes up at a conference. That's because it's like being a kid at a party, when it's your turn to swing the plastic orange bat, and suddenly everyone's watching. On these occasions, I sometimes have to remind myself to be a good party guest and take a good swing at the piñata. But when my turn ends, I have to laugh and let other people have the bat. Similarly, a good conference attendee takes a swing at the intellectual challenge at hand, but then lets others have the microphone and a share of the question-and-answer period. A good guest helps keep the energy up but also helps make sure everybody gets a chance to play.

Poster Sessions and Coffee Breaks

Now, a poster by itself is a weak marketing tool. In today's interactive animated 3-D video world, a poster by itself has almost zero emotional impact. There's no sound, no action. Maybe, if you could put up twelve coordinated posters with flashing lights, you could almost begin to make an impression.

These limitations aside, I suppose the best poster is one that doesn't make the viewer work too hard. As Jennifer Wiseman, Project Scientist for the Hubble Space Telescope put it, "A poster works great if the title tells you the result. So if someone only sees the title and your name they can walk away with something."

Another suggestion I have comes from the writings of Edward Tufte .[7] Often, scientists make posters that are essentially a regular grid of boxes or rectangles. Even if you don't have this layout in mind when you begin, the software we use to make posters often tries to push us into this straitjacket. Given only a few minutes to coach us, I believe Tufte would first insist we turn off the borders on our boxes and erase our grid lines whenever possible, in order to maximize the data-to-ink ratio on our posters.

Tufte would also tell us that our posters must break the tyrannical symmetries of the box and the grid. There are a thousand ways to do so, like using lines and arrows that gleefully puncture the walls of our charts and graphs. Or we can toss images with irregular shapes in among our tidy data points, or choose an overall layout with a joyous visual lilt. There are many more good poster design tips to be found in Tufte's books—far too many to capture here.

Though your poster itself may have limited marketing power, a poster *session* or any coffee break at a conference can be a wonderful marketing opportunity. At a poster session or coffee break, you will a find a large group of uncomfortable scientists waiting around wishing someone were paying attention to them. You can instantly make them your acquaintances or maybe even friends by helping them through this stressful time and showing interest in their work. It's a time for thinking about your marketing funnel (chapter 4): reaching out to new people to draw them into the funnel and staying in touch with people you already know, pulling them down the path to collaborator and then advocate.

Now, when you're standing in front of your poster, there are so many ways to mess up. Here are some examples of what not to say when someone asks you about your poster. I jotted down these bloopers at the last winter's meeting of the American Astronomical Society.

MJK: Hi! Tell me about your poster.

STUDENT: Well, I made it in PowerPoint, and then I printed it out at Kinkos. I almost lost it on the plane.

MJK: Hello. My name is Marc. Tell me about your poster!

STUDENT: Umm, well, the top of the poster is the basic equations. The bottom is a figure from my paper.

MJK: Hey, great to see you again. Tell me about your poster!

STUDENT: Yeah. See, this work isn't really done yet. My advisor says I'm supposed to have it ready by March.

Let's talk about what's going on here. On the one hand, these students are practicing one element of good salesmanship—they are being humble. Their reactions show a kind of endearing humility that might help them win friends in a different setting.

The first problem is that the students' reactions seem calibrated to the wrong kind of interaction—maybe a student-to-student interaction, but not a professional exchange. I walked up and started talking *business*, and they failed to match my mode of communication. The students in this example have failed at code switching, the technique we talked about in the chapter on sales. I wish I had some specific advice to give you about what code you need to switch to at your professional meetings. But it's different in every field and it's even slightly different for every person you meet.

But, here's the second problem with the students' reactions—a problem I think I can address without understanding the culture of any particular subfield. The statements made by the students may have sounded humble, but they were actually self-centered; they were about the poster's owner and his or her misfortunes. When someone drops by your poster, you have only a few seconds to answer this person's primary question: What's In It For Me? (WIIFM). In other words, when you're standing by your poster, what matters are your customer's dreams and misfortunes, not yours. At very least, you owe it to your customer to be engaging.

One good approach is to start by talking about something your customer already knows. Don't forget to mention the brand name of your research idea. Then find a way to start a conversation about your research topic.

MJK: Hi! Tell me about your poster.

STUDENT: Well, you know the new South African survey of galaxy clusters, right? This is a new model for the collisional evolution of these clusters, called "Z-SLAM." So what do you think happens when a spiral galaxy collides with an eliptical galaxy?

If you start by mostly asking questions, rather than stating facts, you're more likely to hold someone's interest and get a conversation going. That's how to find where your interests overlap, and thereby discover a way to answer your customer's WIIFM. As it says in the groundbreaking marketing book *Cluetrain Manifesto*, markets are conversations. And a poster session is a good place to start one.

The Most Important Part of a Meeting

Derek Sivers is a modern music-business success story. He started out as a dreamer, like many musicians, wanting to be a rock star. He went to Berklee College of Music, and he started a band called Hit Me, which garnered some airplay on college radio stations. He sold his CDs at shows and on his website.

Sivers also started selling CDs on his website for other musicians he knew who maybe weren't so computer literate. He started getting these calls: "Hey man, my friend Dave said you could sell my CD."[8] So, in 1998, he started a new site called CDbaby.com to sell CDs for about twenty of his friends. As you might have anticipated, CD Baby quickly became the largest seller of independently produced music on the web, representing over 150,000 bands. In 2008, Sivers sold CD Baby to Disc Makers for $22 million.

Sivers bequeathed the proceeds from the sale to a charitable trust for music education; he made an arrangement that provides him with 5 percent per year of the company's sale price in income. But long before he became a philanthropist, Sivers had already made a good impression on his fans. When NPR interviewed Sivers in 2004, one singer/songwriter commented in the interview, "Derek Sivers is really sincerely driven by wanting to help people."

Sivers writes a blog about marketing and the music business (Sivers. org/blog), which I read fanatically. And once a year, during the annual winter meeting of the American Astronomical Society, I always send out to my colleagues a link to my favorite Derek Sivers essay about how to get the most out of attending a conference.[9] And year after year, I get a stack of grateful e-mails back telling me how useful this essay is. Maybe they are just trying to butter me up, but I doubt it—you know how hard it is to please a critical bunch of scientists.

I've already covered much of the advice in the essay in the chapter on relationship building. "It's about them, not you." "How can you help them?" "Do not push your crap on someone who isn't asking for it." The advice works for musicians and scientists alike. But there's another crucial tip in the essay that I haven't gotten to yet.

Sivers says the most important part of a meeting is the follow-up. Scientific meetings are great places to schmooze. But we often go to meetings, meet new people, and then ignore them until maybe we run into them at the next meeting. Sivers reminds us that the reason we go to meetings is to build real, lasting, working relationships. And most people don't have time to develop these relationships with you at the meeting itself. The real business happens after the meeting is over.

A conference is a frantic time for everyone: you're busy worrying about your presentation, your transportation, and so on, and so is everyone else. In all the noise and confusion, the people you meet can easily forget you and lose whatever stuff you've given them, even if they don't mean to. Most of the time, the only chance you have to launch a real relationship with someone at a meeting is if you take down their contact information, type it into your Blackberry or computer right away, and then follow up with them later. And to start following up, all it takes is an e-mail saying "Hi, I enjoyed meeting you today." I probed Sivers for more details, and he said that first you should send "just a tiny e-mail giving them your contact info, and saying, 'See you around the conference. But let's talk again once we're home and settled.' The main purpose is for them to mentally link your online and offline personas. So when you do the real follow-up a week or two later, that connection has already been established."

Then, the challenging step to remember is to follow up again the next week when everyone's returned home from the meeting, feeling energized and ready to get back to work. You have to e-mail or call and try to connect with that person again. It's easiest to forget to take this step because now you are back in your office too, without the trappings of the conference to remind you that you are supposed to be in networking mode. But Sivers says you'd better not forget. As he puts it, "It's ALL about the follow-up. It's ONLY about the follow-up."

CHAPTER ELEVEN

Giving Talks

 It's easy to get advice from scientists about how to give a good presentation. Everyone seems to have his favorite tips and tricks. "Tell a story." "Tell 'em what you're gonna tell 'em, tell 'em, then tell 'em what you told 'em." "Start with a joke." "Absolutely do not start with a joke."

The advice seems contradictory and mysterious. But once I began studying marketing, the tips and the advice that seemed conflicting at first began to make sense in a new way. What I'm gong to try to do in this chapter is go through some basic pieces of advice on how to give a talk, reinterpreting them through the lens of marketing. Then I'm going to suggest a way to put them all together. I'll show you my favorite formula for a good scientific talk, a formula based on the movie *Star Wars IV: A New Hope* (a.k.a. the original *Star Wars* movie).

Professional Football

My first piece of advice comes from the National Speakers Association (NSA), a club for people who give lots of talks.[1] The folks at the NSA like to say, think of how much a professional football player practices before he steps out onto the field—that's how much you should practice if you're going to be a professional speaker. To become a full member of the NSA, you need to give twenty talks in one year. I leave it up to you to reflect on how you spend your time and decide whether you might be in this category.

I'm sure you've heard this obligatory advice about practicing your talk before. But maybe you thought that this advice was just for students. My informal survey tells me that successful scientists at all levels really do practice their talks, often and out loud. For example, when I asked Nobel laureate John Mather for tips on how to give a good talk, the first thing he said was: "Really do rehearse! It took me a long time to realize I couldn't just do it spontaneously and get it right."

Text on Your Slides

Giving an effective PowerPoint presentation may be an art, but it is not a mystery. Have you seen Al Gore's movie *An Inconvenient Truth*? The whole movie is simply a PowerPoint talk, slightly expanded with some extra video footage. Some people have argued over whether Gore might have done a disservice to the environmental cause via this movie by branding global warming as a white-male concern. But whether or not you agree with this criticism, there is no mistaking the impeccable presentation skills on display in the movie. Gore's PowerPoint talk won him a Nobel prize.

In *An Inconvenient Truth*, Gore used virtually no text on his slides except to label things. Here's why: a listener cannot simultaneously concentrate on your spoken words and your written words at the same time. Text on a PowerPoint slide only distracts you from what the speaker is saying.

Many inexperienced speakers write the words they want to say onto their slides instead of following that first piece of advice (really do practice). Of course, you don't want to do that. It's better to practice your talk enough times so that you don't need text on your slides to remind you what to say. If you can't get by without a reminder, you can use the "notes" feature of PowerPoint. (All this is common presentation advice that I am merely echoing, for the most part.)

However, I want to point out that text on a slide is not always a bad thing. In a foreign country, where the audience is mostly not made up of native English speakers, sometimes your audience will find it easier to read your slides than to understand your speech, which might sound faster and more idiomatic to them than the English they are used to practicing. So when I go to meetings in Japan, for example, I fire up PowerPoint while I'm on the airplane and add text to my slides everywhere I can.

Another time when text or equations on slides might be okay is when you get to the later part of you talk—the part I like to call the action scene. In this part of the talk, it might be better if your whole audience *doesn't* follow you all that perfectly. I'll explain this in a bit.

When You First Take the Stage

One question I've asked many scientists over the years is this: When you give a talk, how should you begin? The answers I've gotten seem to be contradictory. Some people told me "start out with a joke" or "don't take yourself too seriously." Others told me that it is important to establish your dominance on the stage immediately.

When I started working on this book, and asking communications professionals like Randall Larsen and Robert Thompson (see chapter 13) how they give talks, I learned a different point of view. What I learned was that the best way to start a talk is to use the sales technique of *positioning* we discussed in chapter 3—to acknowledge and then work against the audience's preconceptions of you. The best way to start a talk is different for each person with each audience; that's why the advice seemed contradictory.

I might have to break a few taboos here in order to explain what I mean. When you stand up to speak, your audience doesn't know you yet. It's human nature for them to start out by stereotyping you. Fortunately, most intelligent people will move past the stereotype in a few moments, as they get to know you—if you help them out. Yes, I am talking about gender stereotypes, racial stereotypes, stereotypes based on your age, your height, your mannerisms—everything.

In other words, if you're a man like me with graying hair, and someone introduces you as a fancy PhD highfalutin professor, the audience may instinctively stereotype you as someone pompous, snobby, and boring. So your first move on stage should probably be to disarm them with a self-deprecating remark, maybe a wisecrack. You might want to make a comment about beer or football to show them you are one of the guys, or tell a story about your children to soften your image.

However, if you are a short woman or person with a high-pitched voice, you might find yourself facing a different kind of stereotype. The right first move for you might be to launch right into a serious, commanding exposition of the professional topic at hand—to establish your dominance on the stage. You will have to judge for yourself what to do each time you take the stage. Each audience is different.

It may seem unfair to be put in the position of reacting to such stereotypes. But I know I make these same kinds of snap judgments when I meet a new person or see a new person on stage. For the first few seconds I am a victim of my synapses—and I imagine you are too at some level.[2]

Sometimes other people's stereotypes work for us, sometimes against us. But I think the best thing to do when we encouter this flaw in human nature, as we do at the beginning of a talk, is recognize it and help each other move past it. It's good to admit that everyone is human and a little imperfect, and tell each other that that's okay. When we recognize and accept one another's true, imperfect selves, we begin to feel a little bit closer to each other. That's the beginning of a real relationship—the kind that good marketing is all about.

Telling a Story

Clearly, telling a story is a crucial marketing skill; we talked about it earlier in the chapter about sales. And you'll often hear the advice that when you give a talk you should tell a story. John Mather told me, "The audience will enjoy an elementary story well told much better than a complicated story they don't understand. Tell them the story of Goldilocks and the Three Bears, do it well, and they will like it." But it took me some time to figure out exactly how you can tell a story in a scientific presentation, a format that should include a certain amount of technical rigor.

One way I've found to incoporate storytelling into a scientific presentation is to turn my entire talk into a kind of epic adventure. The framework for giving talks that I'm going to present below works this way. It's based on a template that Hollywood uses to lay out the structure for an entire two-hour film. I'll describe this notion in a bit. But I think John's advice refers partly to another rhetorical device that I want to talk about first: each scientist needs to find a short, unique story about her own life experiences to insert into her talks.

Finding a little story about ourselves to tell doesn't sound too hard. As I mentioned earlier, in chapter 3, a story is more or less a sequence of events, with occasional pauses for reflection. We all live in a story, the story of our lives. So we have plenty of events to string together and reflect on. But often, here's what you get when you ask scientists for the story of their work:

I won a grant. I got some data. I fit a model to it. I wrote up a paper. I submitted it and I got back a referee report.

That may well be true—and it's certainly a sequence of events. But it's the same story that every other scientist has. If you want to be entertaining, you'll need to tell a story about your work that's unique.

A robust way to find a unique story to tell is to zoom out. Start earlier in your life—and then leap faster through the action. For example, you might explain what compelled you to get into the field, and say what caught your attention about this particular problem. Then describe some of the obstacles you faced and the people you met along the way. Here's how that might look:

> *When I was in graduate school, I learned that yeast was a good model for human cells. So I started working on yeast, looking for these things called "transposons" that jump from place to place on the gene. Back then, everyone considered the part of the genome I was looking at to be "junk DNA." People were looking at me like I was crazy, which was discouraging. But at one point, I realized . . .*

What really sells this story are the obstacles and the emotions: people were looking at me like I was crazy! I'm sure you've had at least one person tell you you were an idiot during your career. And you forged ahead anyway. That's fun for your audience to hear about.

Once you have dug up an interesting story like this from your past, you can tell it over and over again, each time you give a talk. It's natural to be shy about repeating a story like this over and over, but it's perfectly fine to do so. Even if you become a big celebrity, most people will still not know your stories by heart. And anyway, I'm confident that the storytelling medium is so powerful that people will be happy to hear your story again, as long as it's your unique story. I've gotten so used to retelling my stories at this point, it's almost second nature. But it keeps working. Sometimes I picture myself as a kind of science troubadour, roving around, telling my stories in every town square.

Pitching Your Signature Research Idea

Sometimes you'll hear the following suggestion about how to give a presentation: Tell 'em what you're gonna tell 'em. Tell 'em. Then tell 'em what you told 'em. That's another puzzling piece of advice. Does this suggestion really mean you should only write one third of a talk, but repeat it three times? No, of course not.

The way I view a scientific talk—and any scientific marketing tool—is that it's a means to promote and brand a new research idea: a Signature Research Idea (SRI) or a potential SRI. Most of the time, that means a new research tool (like a new experimental technique or numerical model) or research question (like some puzzling new data or an unproven hypothesis). As we discussed earlier, your SRI should probably have some kind of catchy buzzword name that sounds new to your audience. I think what the "tell 'em" advice is about is the need to mention your new SRI at least three times so people can remember it.

When you hear a commercial for a carpet store on the radio, you may have noticed that during the commercial they tell you the phone number for the store three times. That's because people generally require several exposures to a piece of information before they can remember it. So I recommend mentioning the name of your SRI at least twice during your talk—then find a way to mention it again during the question-and-answer period.

Show a Key Figure Three Times

Mentioning the name of your SRI three times is a good way to help your audience remember it. But this approach works even better if you show an image each time to go with it. When you communicate a message using pictures as well as words, people remember it better; that's called the *picture superiority effect*.[3] So I like to pick one figure that I want everyone in the audience to remember. Then I present this one figure three times during each talk to go with the SRI I'm promoting.

I think the ideal kind of figure to show over and over is a Jenny Craig figure. As you may remember from the chapter on the proposals and figures, a Jenny Craig figure is one that vividly illustrates the impact of your work or proposed work. It shows the state of the art before your research and contrasts it with the state of the art after your research. You can use it to illustrate the predicted benefits of proposed research, or you can use it just as well to show off research you have already completed.

The first time you present the figure, you might want to show only the "before" part of the figure. Ideally, this part of the figure, like a picture of an overweight person, pre-diet, will illustrate the problem that needs to be solved. The second time you show it, you can show how your work changes the situation. Maybe you add some new data points that you collected from your experiment, or a new theoretical curve that fits the data better. Them show it one more time, maybe to help you transition back to the main subject of the talk, or maybe in your conclusion. And it's probably good to leave it up on the screen during the question-and-answer period, to let it sink deeper into people's minds.

Start Simple—Then Step on the Gas

This piece of advice particularly sticks in my craw, because I think I lost a job opportunity once because I failed to heed it. One time when I was applying for jobs, I was on a short list for a faculty position at a university whose name I shall not mention, and I gave what I thought was a flawless job talk. It was well rehearsed and crystal clear. I could feel that the audience was with me all the way from my disarming opening remarks to my final "Thank you." They were hanging on my every word—and their questions after the talk showed me that they had understood my key points.

A few weeks later I found out that I didn't get the job. I had a friend on the faculty, so I asked him what had happened, and where had I gone wrong. He told me that a few people on the hiring committee had complained that they "understood your whole talk." To this day, their reasoning remains opaque to me. But apparently this kind of thinking is not uncommon on faculty search committees and elsewhere in academia. Luisa Rebull, a staff scientist at the Infrared Processing and Analysis Center, told me, "I've heard several anecdotal reports of people being regarded as not very good scientists 'because I understood her whole talk.'"

Maybe, when scientists go to a talk, it takes us back to times we spent sitting in a classroom, in high school or college. We were often bored during classes in high school—a bad memory. Then we went to college and finally met teachers who challenged us, and we no longer understood everything in class the first time we heard it. That was an exciting time—a good memory. Now, when we attend a talk, we want to relive those exciting moments of being challenged for the first time. It's just frustrating, for speakers and job applicants, that this audience desire can run counter to the values of clear communication.

In any case, to please your scientific audiences, I suggest that you plan to make the first half of your scientific talks mostly accessible to anyone who is scientifically literate. Then, roughly halfway through, step on the gas and focus on the experts in the audience for a few minutes. Here is where we diverge from Al Gore's presentation style in *An Inconvenient Truth*. Spend a few slides getting technical, and showing off your specialized knowledge. Show the equations. Use the jargon. Talk fast. Then slow down and bring everybody back in for the last few slides.

I like to think of this fast-talking, technical part of the talk as analogous to the action scene near the end of a Hollywood movie, when the characters are racing around to stop the bad guys before their evil plan hatches. I'll talk more about this analogy in a bit.

Depth Spikes

Many audience members will probably be perfectly happy with start-slow-then-step-on-the-gas approach. But others may start to get antsy during the first half of your talk, if you are reviewing the foundations of a field in which they consider themselves to be the local experts. During this time, you can interject occasional "depth spikes" into your talk in order to pacify these local experts and other cranky people, and remind your audience you are a technical pundit.

A *depth spike* is a brief side comment addressed to the experts in the room—maybe translating your words into the jargon of your subfield. I learned this idea from my friend David Goldhaber, who is a professor at Stanford. Here's an example:

> For those of you who work on photochemistry, this trend is equivalent to the Stark-Einstein law.

The Question-and-Answer Period

When you are done with your slides and the time comes for the question-and-answer period, it's tempting to stop and look at audience for approval, and then regress back to being a student under examination. You have just sung your part and you are feeling vulnerable. You want someone to pat you on the head tell you you're okay.

But nobody pats you on the head and tells you what a good job you did. The questions start coming, and now you don't have a script to follow. So you start quaking in your boots, feeling insecure. The sharks in the audience smell blood and move in for the kill, hogging the air time, grandstanding, and promoting their own brands and ideas.

Or course, there's a better way to handle the situation. You can turn the question period into an extension of your talk time. It can be a time to recite the name of your brand again, reiterate its strengths, and recite some more memorable facts about it.

To play this game, first, you'll need to prepare three or four memorable points to make about your SRI—fun factlets, or simply points about how your SRI is new and different from what's come before. These can be points right from your conclusion slide or from the conclusions or abstracts of your papers. They can be sound bites from your press release.

Then when people ask you questions, you can work these points into the answers. Sometimes the questions will be amiable—and you'll simply answer them. But before your take the next question, you might want to add: "What I want you to remember is . . ." and launch into one of your sound bites. That maneuver is called a "pivot."

If the question is not so constructive—if, for example, it seems to be mostly an advertisement for someone else's brand—you don't have to answer it. But you still have to respond—and you can still use the opportunity to promote your work. For example, you can say, "That's an interesting topic; thanks for bringing it up. But what I want to tell you about today is . . . ." In other words, you can pivot back to one of your prepared points about your brand. The audience will be behind you if the question was clearly a selfish one.

Star Wars

Part of the experience of being a songwriter is always chasing the formula for a hit song. Of course there is no one formula that covers all good songs. But a few minutes of actively listening to the radio will show you that many songs have similar structures, often the pattern verse-chorus-verse-chorus-bridge-chorus. With a few hours of active listening, you'll notice many other common patterns, as well as a sense of when a particular song is deviating from the standard patterns.

Similarly, while listening to many colloquium speakers over the years, I've been chasing the formula for the perfect colloquium talk. Just as in songwriting, I expect that there is no single formula that all good talks must conform to. But I think I have found one that's pretty compelling and can be adapted for a public presentation as well. Let me to run it by you.

My formula is based on what's called the Three-Act Structure or the Hollywood Story Arc. The Three-Act Structure is a kind of universal storytelling pattern that a large fraction of Hollywood films use as a template. It was canonized by screenwriter Syd Field in his 1979 book *Screenplay*; scientist-turned-filmmaker Randy Olson also describes it in his book *Don't Be Such a Scientist*.[4] The pattern is related to what Joseph Campbell calls "the Hero's Journey"; its roots reach far back into ancient mythology from all around the world.[5]

To explain the formula, I'll use as an example *Star Wars IV: A New Hope*, a film that Field analyzes in great detail. (Back then it was just called *Star Wars*.) I'll describe it using the language of archetypes (see chapter 6).

First, an opening panorama sets the tone of the film: the camera pans across a tremendously long and complicated-looking spaceship. Then we are dropped into the middle of the action (the literary term is *in media res*) and we watch a confrontation between Darth Vader and Princess Leia. We don't quite understand everything that's going on at this point, but we certainly get the idea that what's happening is of great importance to the characters on screen.

Next the action slows down to a pace we can all easily follow. We meet the main character, Luke, who at this point is the archetypal innocent. Luke has an encounter with a strange new culture and realm: the world of space battles and Jedi Knights, R2-D2, and Obi-Wan Kenobi. Then he suffers a crisis: he discovers that the farm where he lives with his aunt and uncle has been destroyed and his aunt and uncle have been killed. Luke is forced to leave his comfortable home and embark upon a quest to look for answers, with the help of Obi-Wan (a sage).

Luke meets a villain (the evil ruler Darth Vader) and suffers severe trials that nearly discourage him. But he finds a new source of strength ("Use the Force, Luke!") and decides to resume the quest. He heads off with a new goal: to vanquish the enemy by destroying its new weapon, the Death Star. There is an action scene (a chase through the canyons of the Death Star). The protagonist (Luke) destroys the Death Star and finally emerges as a victorious hero thanks to his new source of strength (the Force).

Now let me try to show how the storytelling structure of *Star Wars* can apply to a forty-five-minute scientific colloquium talk. I think you'll see that this formula can incorporate all of the tips we just discussed, and provide a graceful storyline to your talk, ending with you and the new SRI you are branding now cast as humble-but-victorious heroes. Here's how it goes:

When you open your talk, you'll want to start by surprising the audience a bit by doing something to work against their preconceptions: positioning yourself as we discussed above. You'll also want to show some visually exciting Beautiful Butterfly figures to get the whole audience in the mood. This part of your talk is like the opening scene of *Star Wars IV: A New Hope*, where the camera pans across the shockingly huge and detailed Imperial Star Destroyer. That scene entranced audiences with its visual impact, and it put to rest any preconceptions that the audience might have had that the quality of the special effects in the movie would be weak, as they were in many science fiction movies of the 1970s.

Next, it is time to briefly address the experts in the room and give some recognition to the other people in the field. A good way to do that is to show a Family Portrait figure, a figure that shows the work of many previous groups on one viewgraph. It's okay if you speak over the heads of some of your audience for a minute here while you are addressing the experts. This section of your talk corresponds to the scenes in *Star Wars* where the viewer is dropped in the middle of the action, and doesn't quite understand all that's going on.

Then, you zoom out and tell your personal story of how you first got involved in the field, like the innocent Luke's first encounter with the curious droids, R2-D2 and C-3PO. Here is where you innocently mention your previous work in the field, as well as the steps you took that didn't quite satisfy you but only intrigued you further. You'll want the whole room to be following along at this point, but you can use some depth spikes to show your expertise and thus stay connected to the experts in the room.

Then, still following the *Star Wars* script, fate hands you a blow and your life is uprooted. You realize that the latest theory does not fit the latest data! Or perhaps you find there is new region of parameter space that has become crucial to explore. For you, in your talk, this is like the moment when Luke discovers that his home has been destroyed by the evil empire. You come to grips with this turn of events, and you embark upon a quest to answer your research question or use your research tool, like Luke setting off with Obi-Wan.

Naturally, while you are on your quest you meet obstacles. So now you explain why the problem you have chosen is hard. Perhaps the experiment is delicate or the calculation is challenging. A good way to do that is to show the "before" part of your Jenny Craig figure. Luke, of course, meets many challenges during the corresponding part of the movie— Darth Vader and the Storm Troopers, for example.

Now you introduce your SRI, a new research tool or research question that you have provided with a catchy name. This new idea might be able to solve the problem, explain the discrepancy, explore the parameter space, and so on. Your SRI is like the Force, which Luke must use in order to destroy the Death Star.

Next comes an action scene, where you step on the gas, strut your technical stuff, and in rapid fire, explain the juicy details of your new idea at a level only a small handful of top experts in your field would completely understand. This part of your talk corresponds to the chase scene near the end of *Star Wars IV: A New Hope* where Luke speeds through the metallic canyons of the Death Star, chased by Tie Fighters. To appreciate this scene, the audience does not need to follow where every laser beam went or exactly which spaceship exploded. But they do need to get the sense of excitement and challenge and risk. In your talk, this scene should last long enough to illustrate the depth and complexity of your thinking and your process, but not so long that the audience members who can't follow all the details begin to lose interest.

Luke then emerges from the canyon having miraculously hit his target, using the Force. You emerge from your technical fireworks show having solved the problem using your new SRI. The movie screen fills with a shot of the Death Star exploding. You show the second half of your Jenny Craig figure, vividly demonstrating what new advance you have brought to the field.

Table 11-1   A *Star Wars* Approach to Giving a Talk

Star Wars IV: A New Hope	A Scientific Colloquium Talk
The camera pans around an intricate spaceship floating in space to set the mood.	You show some Beautiful Butterfly figures (pretty images) to set the mood.
There is some serious action involving spaceships and droids, which likely mystifies you at this point, unless you have already seen the movie.	You show some Family Portrait figures, which summarize the work of many research groups on one chart, speaking to the experts in the room and illustrating how serious this field is.
Innocent Luke meets some curious robots.	You zoom out and tell how you innocently got interested in the field.
Luke finds that his aunt and uncle have been killed and his home has been burned down.	You explain to everyone how you found that the prevailing theory fails to fit the data. You show the "before" half of your Jenny Craig figure.
Luke sets out on a quest to find the answers.	You tell how you naively set out on a quest to find the answers.
Luke meets the evil ruler Darth Vader.	You meet some challenges. This problem is hard!
Luke escapes and returns with the Force, a crazy new idea that just might work.	You invent/discover/hypothesize a new Signature Research Idea (SRI) that might be a way past the roadblocks.
Action scene: a race through the canyons of the Death Star chased by Tie Fighters. It's exciting, though you can't follow every laser beam.	Action scene: you zoom through the technical details, rapid fire, now addressing only the experts, speaking over the heads of most of the audience. Some frown and look puzzled.
Luke blows up the Death Star using the Force, but the evil Darth Vader gets away. Luke is awarded the medal of a hero.	You show both halves of your Jenny Craig figure. Turns out, your new SRI works and changes the course of science, though important problems remain. Your new SRI, and you by implication, are heroes.
The credits roll: May the Force Be with You!	You show your Jenny Craig figure one more time at the end and leave it up on the screen during questions. When you answer them, you mention your SRI.

But Darth Vader escapes. And just as Darth Vader escapes, leaving problems to be solved in a future movie, you find that there are still scientific puzzles that remain for you to solve, and further work to be done. As we discussed earlier, whenever you unlock a mystery of the universe, you must quickly replace it with another locked mystery.

Table 11-1 sums up this script. Of course, the form has limitations; it's probably hard to adapt for a five-minute talk, for example. And it might not help you when you are reviewing someone else's work at journal club or grand rounds. But I've found I can use it, or some version of it, in the majority of talks I'm called on to give these days. In any case, whatever form you choose for your next talk: may the Force be with you.

CHAPTER TWELVE

Internet and E-mail Marketing

Veronica McGregor is a science writer at NASA's Jet Propulsion Laboratory (JPL). One day, the team she was working with calculated that their Mars probe was going to touch down during Memorial Day weekend. So McGregor's manager asked her to set up a Twitter feed to send out updates about the project, hoping to keep the news flowing over the holiday.

McGregor was new to Twitter's 140-character format, and—a first for her—she struggled to communicate. But soon she got the hang of it. "I would write one day 'the space craft has traveled' and it took up half my space, so then I started to cut out words. Finally I cut out the spacecraft and said, 'I.'"[1]

Then a strange thing happened. The Twitter page (twitter.com/MARSPHOENIX) became more than a news feed. By dropping "the spacecraft" and saying "I" instead, McGregor became the *voice* of the Mars Phoenix Lander.

People started asking McGregor—that is, the Lander—personal questions. "Are you male or female?" And the questions kept coming and coming. She said, "It sounded like a Vegas slot machine. My computer was just going ding ding ding." Soon, stories about the Twitter feed appeared in the *New York Times*, *Wired Science*, and ABC News. The Phoenix Lander mission is over now, but McGregor still has more than 88,000 Twitter followers.

Everyone knows by now the remarkable power of Twitter for communicating to large audiences. A former aide to George W. Bush has even argued in the *Christian Science Monitor* that Twitter deserves to win a Nobel Peace Prize.[2] The story above illustrates this power, but it also shows a glimpse of some of the aspects of Internet marketing that puzzled traditional marketers; the Internet rewards trial and error, creativity, and a personal touch. Throughout this book, we've been discussing how the advent of the Internet reshaped marketing philosophy. But now let's take a look in more detail at some of the many Internet marketing tools scientists can use, and we'll keep an eye out for the creative ideas and personal touches that creative marketers have used to find the human dimension of this new medium.

Your Research Website

Young scientists sometimes tell me that they don't bother working on their websites because nobody ever looks at them. Then I tell them about the last time I sat on a hiring committee, and how the group of us all sat around Googling the candidates. Or I tell them about the time when I served on a review panel and we were deadlocked, and we broke the tie with information from the proposer's website. That usually changes their minds and sends them rushing to their computers, html manual in hand.

Of course, it's reasonable to be dissatisfied with an ordinary website these days. The real magic of the Web is clearly in the interactive sites, like Facebook, Twitter, and Foursquare. These websites turn everyone into a participant—consuming content but also creating it. The phenomenal new dominance of interactive websites across the Internet is often referred to as Web 2.0. In contrast to Web 2.0 sites, the websites most scientists can build easily at their home institutions are generally static, noninteractive, and therefore intrinsically boring.

But even if you are a Facebook fiend, a Gowalla guru, and a Tumblr plumber, I still think you have to have some kind of dedicated research website even if it is just a noninteractive Web 1.0 site, if that's what your home institution offers. You could think of it as a sign outside your shop. Unlike the salesperson inside, it doesn't talk with you. But it can still help your customers find the place.

Your home institution likely has some tools for helping you set up a website; that's the best place to start with your home page construction project. Also, the folks on my Facebook page highly endorsed Wordpress. org as a tool for building websites. Wordpress can help you construct a simple noninteractive website and also allow you to easily incorporate blogs and other interactive features.

For bigger web projects—those with an actual budget—I suggest 99designs.com, a site that helps you hire a web designer, cheap. You post information about the site you want built, and you name the price you are willing to pay. Then designers submit designs, and you choose the one you like the best. You can even have the designers create a logo for you. And if you don't like any of the designs that are submitted, you get your money back.

Now let's put ourselves in our customers' shoes and consider what we should put on our home pages. First of all, surfing the Internet can be a rewarding experience if it's easy to find what you want, but a frustrating experience if it takes too many clicks to get there. As Robert Naeye, editor in chief of *Sky and Telescope* magazine said to me, "People on the Internet have a short attention span. If you have to click to go to a second page, I don't usually click to go to a second page." That's one customer preference to keep in mind.

And of course, there are many ways to make a website annoying. I'm sure you have already developed a taste in websites that includes a sensibility about these things. For example, I'm sure you have heard that you should mostly stick to dark text on a plain, light-colored background, and never make your text blink. And all the usual advice about typesetting applies to websites as well as the printed word. For example, it's nice to leave lots of white space and stick to columns of text with no more than about 60–80 characters per line.

To get some more hints, I talked to someone who has spent many hours scanning science websites, science journalist Genevive Bjorn. She gave us scientists low marks for visual clarity. "What I often see on scientists' websites is too much visual noise," she said. "Website elements that some scientists think are fine—such as long scrolling lists, multiple text colors and fonts, or patterned backgrounds—often don't translate well to non-scientists and may even become an annoying obstacle for someone like me who doesn't have a lot of time to spend on each page."

Now, let's think about how our websites can help our colleagues. As I've argued, websites can be good ways to offer free samples of our work; we can post products there to be useful to our colleagues—software, catalogs, figures. If you create a page that's full of useful resources, you may suddenly find everyone in your field is posting links to your page and spreading the good word about you.

Our websites can also help us develop the brand that is you. Maybe you are a fastidious experimentalist or you are known for you out-of-the-box creative theories. Perhaps you are you known as a leader of large projects and of other scientists. You may have some archetype you aim to project, like the ruler, outlaw, creator, or hero. A website can paint any of these pictures.

For example, if you are known for your numerical modeling (or you wish to be so known), you might want to post links to supercomputing centers, a picture of one of your latest simulations, and maybe even some links to classic computer-science puzzles. If your expertise is fieldwork or lab work, you might want to post a picture of yourself in your pith helmet or your lab coat. I've found it handy to post both a thumbnail and a large, high-quality photo—one that someone from the press can easily grab and publish.

A website can easily be used to project your archetype, using symbols, design, and color choices. For example, symbols like flags, rockets, and military camouflage seem to go with the hero archetype: Bright, smooth primary colors, or gears and tools, can indicate a creator. A slick, professional-looking design and a generic photo of you in a suit next to some marble pillars would spell ruler. Flames, black, leather, chains—that's the outlaw, or maybe the orphan. I suppose the challenge here is to be subtle about it—we aren't selling detergent or motorcycles here—and somehow gently to merge one or two symbols of our archetypes with the symbols of our science. Maybe for guidance, we can look at the websites of some successful archetypal scientists like Stephen Hawking, Sally Ride, or J. Craig Venter.

It may make you feel uneasy to post your own biography or CV. But it's actually helpful to your colleagues to have this information handy when they need it for proposals and reports and so on. Likewise, you'll want to post your contact information to help people find you. I also post links to the websites of my students and postdocs to help people find them.

Finally, a key audience for your website is the press. With a little effort, we can make our websites useful for them as well, as Genevive Bjorn explained to me:

When I visit a scientist's website, I am usually looking for sources, people to interview for articles. I typically need to know their name, what they work on, where they work, and to see links to three to five recent papers. Ideally, at the top of the webpage I see a phone number or an e-mail and a picture so I know who I'm contacting. And the page should be in a simple and easy-to-scan format: white background with black text in a sans-serif font (like Verdana) surrounded by clear section breaks and lots of white space.

Another thing that's helpful is an obvious link to their prior work with the media. This might include previous articles that the scientist has been quoted in, a list of media coverage of their particular work, or links to projects that he or she has participated in for the public. This kind of information shows that the scientist knows how to work with the media. It also sends the signal that they want to talk to the public.

The E-mail Newsletter

Every songwriter subscribes to tip sheets. The tip sheets come once a week, listing recording artists who are looking for songs. Every week I would go through the listings, pick out songs from my catalog that seem to match, and send out a bunch of CDs or e-mails containing those songs to the artists. Then, at first, I would just cross my fingers and hope that one of those seeds would take root—that someone would like one of the songs I sent.

Then came a turning point in my country music career. Once I started studying marketing, I realized that these tip-sheets are not just lists of song requests—they are lists of potential long-term clients. My job, I suddenly realized, was to get to know these potential clients and slowly draw them through my marketing funnel.

When I started this process, not every aspiring country artist had an e-mail address yet. But I started collecting the e-mail addresses of those who did, and adding them to a list. Then I had to make up an excuse to develop relationships with these artists. So I started sending out a monthly newsletter with two or three tips about the music business that might interest them. For example, I might remind them about an upcoming audition for American Idol or I might tell them about my experience with a new music website. It took me a few hours per month to put my newsletter together. I mixed in a joke here, a personal anecdote there.

I never hawked my songs in the newsletter. I kept the content about them—about life as a country singer, trying to make it big. Though, of course, the signature line at the bottom of the e-mail said "songwriter" and had a link to my webpage.

Sometimes, I would send out the e-mail and nobody would reply. Sometimes I would get a few people thanking me for the tips, and then silence. But sometimes, the day after I sent my newsletter, someone would write back with a note of thanks, and by the way, a request for permission to record one of my songs. That's how I've gotten most of my "cuts" so far (a cut is a song recorded by a recording artist).

When songwriters go to Nashville, the thing to do is try to meet with as many music publishers as possible—I've made the rounds. At first I was green and timid, begging for any advice I could get. But after a few years of getting a steady stream of cuts through my tip sheet, some publishers actually started asking me for advice.

So of course this experience led me to examine my scientific life, and to consider whether I could develop a mailing list of potential scientific clients. It seems to me like there are many excuses for a scientist to develop such a list. Some scientists send out monthly or weekly e-mail newsletters about papers published in their field. Others organize a seminar series, giving them a reason to e-mail their whole department once a week. If you have information that *everyone on your list would actually like to know*, you can use an e-mail list to simultaneously keep a long list of potential clients moving down your marketing funnel. You send them a little bit of love in each e-mail.

Regular mass e-mails are wonderful, but even a one-shot mass e-mail is good. For example, when you publish a paper, I think it is good idea to send out an e-mail or snail mail advertising the paper—but only send it to scientists who you are sure will care about the paper. As bioengineering professor Deborah Leckband told me, "It's actually helpful to people because otherwise most of my colleagues are too busy to keep up with the literature. One of my colleagues said the only way he kept up with the literature was by reviewing proposals!" If you are in doubt about who will care about your work, you can send your announcement only to scientists whose papers your paper cites; they will at least take pleasure in seeing their names in the reference list.

Part of the key to my country singer newsletter was that my e-mails all ended with my contact information: my e-mail address, phone number, and website. You can set this up in most e-mail tools, or if you are a Linux person, you can use a "signature" file. It's amazing to me how many scientists do not have e-mail signatures with their contact information; they are missing out on an important marketing opportunity.

Now, we all have to spend some of our time as scientists on community service—tasks that are not directly research, but which otherwise benefit the community. If you have a choice of what service opportunities to pursue, I suggest choosing ones that will give you a chance to send out a mass e-mail every now and then. For example, you could organize a talk series at your home institution, or organize a meeting of your subfield. That way, at regular intervals, you have a good excuse to send a few hundred people a message that they will actually be grateful to get.

Home pages and e-mail newsletters have become enshrined as standard scientist-to-scientist communcations tools, unlikely to offend even the most marketing-phobic senior scientist. Now let us take a look at some tools that are still new to much of the scientific community, or that presently have a reputation among some scientists as public outreach tools. Whatever reputation these new Internet marketing tools may have now, I suspect their influence on science will grow.

YouTube

The summer before the Large Hadron Collider (LHC) came online, a key LHC press officer (Dr. Katie Yurkewicz) went on maternity leave. Kate McAlpine, a science communication intern half a year out of college, stepped in to help. Her main duties were to arrange visits for U.S. journalists to the collider and to accompany the journalists on their visits.

McAlpine also borrowed Yurkewicz's video camera and, in her spare time, made a short video about the Large Hardon Collider. In McAlpine's LHC video, science communication interns break-danced next to the collider's giant magnets, while McAlpine rapped about the beginning of the universe and what we would learn about particle physics when the collider went on line.[3]

McAlpine returned the video camera, left the position, and went to work for another project. Then she posted the video on YouTube in July, a month and half before the collider was scheduled to begin operations. With its fast-paced, catchy music, the video became a phenomenon. It quickly garnered about three million views, and lots of good publicity for the collider and for particle physics in general. By now, her video has had about six million views, more than some rap videos from major record labels.

McAlpine used a basic handi-cam video camera, the kind you might use for family home videos. Kate's friend, Will Barras, created the music on a Mac using Apple's Logic Studio. "The video sat on my computer for a while," Kate told me. "Windows Movie Maker had a tendency to crash after I had a minute's worth of video edited. I had to edit in minute-long chunks and tie them together at the end." But the video's low-budget nature seemed only to boost its success.

Many scientific institutions have started regularly posting videos on YouTube with every press release. These official productions are often serious affairs with talking heads and captions and voice-overs. Such serious scientific videos can be enchanting; I've contributed to a few myself. But, as the example of the Large Hadron Collider video shows, audience can sometimes respond better to the zany, the amateur, and the *real*.

Part of what has made the amateur science video such a successful genre could be that it's a fresh new form. These videos are generally free of the clichés of the professional science videos (droning voice-overs, talking heads). They may even parody these clichés.

But is it possible that the amateur science video is itself starting to become a cliché? Can you still make a viral science video with a one-hundred-dollar camera? Geeked on Goddard Science blogger Daniel Pendick warned me that there are now some minimal production standards even for amateur science videos. "People will overlook marginal video skills, but bad sound will send them running for the exit doors."

I asked McAlpine if she thought the science rap video was all played out. She said, "I don't think so. There are other people out there rapping about science and people who are enjoying these videos. As long as you make something that's a little bit catchy that can also tell you something that's interesting. I think there's even a place for the stuff that makes you cringe too—like my first video or perhaps all my videos. Because it's just people goofing off and enjoying their science."

Web 2.0 and Blogging

Facebook, YouTube, MySpace, Twitter, and Wikipedia are all examples of what's called Web 2.0. At a Web 2.0 site, the person who surfs to your site gets to play a starring role by contributing his or her own text, images, videos, music, and so on. When people can participate in the construction of a website in this way, they feel like they own a piece of it, and they enjoy coming back to look at it, again and again. Web 2.0 is more than just an buzzword. Web 2.0 sites have become vast worldwide communities of happy, participating customers and proselytizers—microcosms of the marketing ideal. Every company, even one hopelessly bound to bricks and mortar, has been touched by this new social and technological phenomenon.

A simple Web 2.0 option that you can implement is a blog where you provide updates, and visitors can leave comments. You can use the tools in Wordpress, or any of several free blogging engines, like Tumblr (www.tumblr.com) or Blogger (www.blogger.com). Visitors to your site will want to keep coming back to check and see if someone has comments on their comments. You only need to post new conversation-starting material maybe once a week, if you have something interesting to say. But the real magic happens when you give people quick feedback on their comments. As McAlpine told me, "I think what people are just catching onto is that it's really all about interaction. If you want to get followers and keep them, you have to talk with them and get back to them when they talk to you."

But the days when starting a blog was a shortcut to fame and fortune are long gone. Current estimates put the number of blogs on the Internet at around 10 million.[4] Even the Federal Government blogs; USA.gov lists more than 100 blogs run by U.S. federal government institutions, including the White House, the Marines, and the Transportation Security Administration (TSA).[5]

If you start a blog now, and you want to generate appreciable traffic, you'll probably need to be representing an entity that's already well known, or you'll need to invite your friends to read it, and you'll have to put in a fair bit of effort before you build any sizeable audience. You'll be competing with a growing army of full-time bloggers who entertain visitors with pictures, videos, and well-researched articles. If you are only representing yourself, and you don't have time to regularly create a lot of new content, you may find Facebook a more rewarding environment for growing your Web 2.0 enterprise (see below).

Social Networking and Microblogging

Recently, I had the deep pleasure of informing a graduate student that he had been selected to receive a prestigious postdoctoral fellowship. I called his office. No answer. I e-mailed. No reply. I sent a message on Facebook, and he called my cell phone ten seconds later.

Social networking websites like Facebook, LinkedIn, and Twitter have emerged as tools that we can use to initiate and maintain professional relationships. The idea is you set up a webpage for yourself by filling out an online form and uploading a photo, and this page becomes part of the social networking website. Then the social network provides a variety of ways for people to find your page based on the information you entered, and to create a connection between their page and yours. Creating connections—that's what marketing is all about.

Part of the power of these tools seems to come from your picture in the top left corner of your page—it touches people to see who they are communicating with. So a Facebook message, for example, while it's still not as good as a phone call or visit, may be more personal than a regular e-mail. Maybe the photograph in the corner of the Facebook message I sent is what appealed to this particular postdoc.

I mentioned Facebook, LinkedIn, and Twitter. Well, each of these social networks has a different character and attracts a different range of users. For example, far fewer people are on Twitter than on Facebook; joining Twitter doesn't much help you catch up with old high school buddies. But people who tweet (sometimes called "tweeps") tend to be folks who are interested in spreading ideas—marketers, like you. Facebook is an especially effective way to communicate with younger scientists, as the story above shows. LinkedIn (www.linkedin.com) is another social networking website, one that is more popular with older professionals and job seekers.

Online social networks have encouraged a phenomenon called "microblogging." That's the practice of sending out short bursts of information and links, limited in length to a string of roughly 140 characters, to a list of recipients. On Facebook and LinkedIn, this string is called an "update." On Twitter it's a "tweet." Whatever platform and terminology you like, there's no need to limit yourself to microblogging on one social network. You can set up your accounts on Twitter, Facebook, and LinkedIn so that when you post updates, or tweets, they go to all three places at once, automatically.

A common mistake many scientists make is posting trivial personal information on Twitter or Facebook every 20 minutes: "I'm taking the dog for a walk, then grilling some cheese for lunch." If you plan to use Facebook or another social networking website for professional marketing, you wouldn't want to do that.

Your posts and tweets only help you professionally if they are entertaining or useful to your colleagues (see the fundamental theorem of marketing). So rather than posting boring updates about your lunch menu, you might want to post links to websites or news articles—and add your commentary. Jokes, questions, challenges, and dares all make good posts. It's also better to limit your posts to one or two a day, so you don't look like you're spending so much time on Facebook that you aren't doing any work.

Another common mistake new microbloggers make is sending out too many self-promotional tweets ("check out my new website!"). The idea of Web 2.0 marketing is not to inform people what you're doing— it's to start a conversation. Twitter guru Chris Brogan (91,000 followers) recommends that you tweet in a 12:1 ratio: for every one post that directly promotes yourself, promote twelve other "good/interesting things."[6]

If you find yourself getting serious about microblogging like Veronica McGregor, you'll probably want to start using a tool like Tweetdeck (www.tweetdeck.com) to schedule your tweets/updates in advance so you don't have to log in every single day to maintain your presence. That's how the pros do it. And once you have established a bit of a following, you might enjoy tracking your "influence" on the web with tools like Klout (www.klout.com) and Peer Index (www.peerindex.net).

Groups and Pages

A feature of some of these tools that many scientists overlook is the ability to use them to set up your own Web 2.0 environment to market your work. For example, if you think of your Facebook profile as a kind of professional webpage, it can become a powerful interface for connecting with your colleagues. You can "friend" lots of colleagues, and then regularly update your status so that your friends get regular notes from you. Facebook allows you the option of sending these updates to your Twitter account as well (as "tweets").

If you chose to do that, you'll have to be extra careful about what you post on your personal page. Besides boring your colleagues with trivia, there are any number of horrible gaffes that can arise if your colleagues can access your personal information on a social networking site. I'm sure you would know better than to claim to be home from work sick, while allowing your friends to post compromising pictures of you at a rave, impersonating John Holdren. But scientists I know have indeed embarrassed themselves by posting irrelevant material during a colleague's talk that they claimed to be paying attention to. Fortunately, Facebook now has a detailed system of privacy settings that apply to all different aspects of your profile. If you create a list of all your professional contacts, you can easily prevent them from seeing just about any private information you might have on Facebook by using the "customize" options and typing in the name of the contact list.

If you want to create a Facebook network of business connections separate from your friends network right from the start, you can start a "page" or "group" for your professional goals. LinkedIn allows you to create similar associations, also called "groups." These groups allow you to conveniently reach out to every group member who's paying attention—a bit like an e-mail newsletter. You just have to make sure you are sending out something that everyone will actually find humble, respectful, and valuable, just like a mass e-mail. And Facebook pages and groups allow everyone in the group to comment on whatever's posted to the group, encouraging lengthy conversations.

I love Facebook groups. I've found the Marketing for Scientists Facebook group to be a fantastic source of ideas and connections. If you think you will get enough participants, I strongly suggest experimenting with starting a Facebook group or page for your subfield, or even for your SRI. Facebook pages also have proven track records. The day after the 2010 midterm elections, Facebook announced that in 74 percent of the U.S. House races that had been decided up to that moment, the winning candidate was the one whose page had the most Facebook fans.[7]

The difference between a fan page and a group is a slightly confusing aspect of Facebook marketing—one that I haven't yet entirely figured out myself. The functional differences are not too hard to spot. You can add members to a group from your list of friends; you can only invite people to "like" your page. A group shows links to your personal page and the pages of people you designate as administrators. But a fan page does not show these links, or any connection to your personal Facebook account. Fan pages are also searchable by Google, while groups are not. And, crucially, a fan page allows you the option of buying precisely targeted advertising from Facebook to promote the page to new potential customers. In short, fan pages seem to support a less committed, less personal interaction; they are for reaching out to people at the mouth of your marketing funnel. Groups are meant for interacting with people further down your funnel.

In general, if you are trying to use social networking for professional goals, and aren't thinking in terms of your marketing funnel, social networking can waste a lot of your time, as can any form of promotion. For example, many scientists make the mistake of spending too much time on the mouth of their funnel, and neglecting to draw people in to more intimate, collaborative relationships. As wonderful as is can be, social networking is still not as real or rewarding as a phone call or an actual face-to-face interaction. So if you find yourself meeting new people on Facebook and you're ready to move them down the marketing funnel, you're probably going to have to move the conversation to the phone or to Starbucks or to someone's office, as soon as that makes sense. I sometimes have to remind myself: the Facebook "friends" who actually *are* my friends are nearly all people I've actually met.

Staying Focused Online

The Internet offers so many opportunities, it's easy to get lost and confused. New Internet communication companies keep popping up, clamoring for our attention, offering new ways to connect with people and new ways to display ourselves. Amid all the confusion, sometimes it can be hard to tell whether I'm actually furthering my goals or simply helping someone else sell advertising. And it can be easy to succumb to a simple urge to self-promote, instead of remembering the point of it all: helping make the world a better place through our science.

There's a great marketing quote from Herb Kelleher, the former CEO of Southwest Airlines, that has helped me make my way through this jungle. It's about branding, and need to stay focused (see chapter 5). Herb liked to teach his employees a mantra to help them keep focused on the Southwest brand. He said the following:

> I can teach you the secret to running this airline in 30 seconds. This is it: Southwest is the low-fare airline. Not *a* low-fare airline. We are *the* low-fare airline. Once you understand that fact, you can make any decision about this company's future as well as I can."
>
> Here's an example. Tracy from marketing comes into your office. She says her surveys indicate that the passengers might enjoy a light entrée on the Houston to Las Vegas flight. All we offer is peanuts, and she thinks a nice chicken Caesar salad would be popular. What do you say?
>
> You say, "Tracy, will adding that chicken Caesar salad make us *the* low-fare airline form Houston to Las Vegas? Because if it doesn't help us become the unchallenged low-fare airline, we're not serving any damn chicken salad."[8]

Kelleher's brand mantra helped him decide whether any new marketing venture was worthwhile. As a scientist, you can use this approach to stay focused on your goals while you surf the web. Southwest is *the* low-fare airline. Maybe you are *the* maverick quantum bioinformaticist. Does this new website help you reach the top of your field by winning new converts to your useful Signature Research Idea? Whatever Internet techniques you chose to use—webpages, e-mail lists, Facebook pages, tweets, blogs, videos, whatever tools the future brings—they should all pass the test of your mantra before you decide to devote a substantial part of your time to them.

For example, Google came out with a new social networking tool this summer called "Google+." Should you start spending half an hour a day on Google+ to market your science? Herb Kelleher might ask you: well, is there anyone yet on Google+ who might want to use your bioinformatics SRI, or help you improve it? Or is it just some damn chicken salad?

Googling Yourself

The other day, I was slated to introduce a visiting seminar speaker, and I didn't know what to say. It was a busy day, and there was no time to prepare; all I could do was sneak off down the hall for a few seconds and Google him using my smartphone. Alas, I found only a long list of out-dated pages, and no material I could use, not a single fun fact. Ideally, I would have been a more conscientious seminar host and done this search in advance. But imagine other, busier, more important people than me looking for information on this speaker: reporters, hiring committees, and so on. They might just give up and move on, and he would never know what opportunities he had missed.

Trying out the range of social networking sites on the Web can be fun and it can lead to good professional relationship building. But often, when you Google a scientist nowadays, you'll pick up a LinkedIn page, a Facebook page, an old home page or two, maybe a blog or Twitter feed, and maybe a page on Nature Network.[9] Half of these links will list no contact information, or maybe just an email address, with no phone number and no CV. Many of them will contain out-of date addresses, or job titles. All these incorrect or redundant pages may augment your Google hits, but they can easily dilute your message and confuse your customers.

So Virginia Gewin, writing in *Nature Jobs*, recommends Googling yourself every now and then—not for vanity, but to look at the results as a busy stranger might.[10] You may want to prune whatever forest of sites might appear, or work on keeping them consistent, informative, and up-to-date. Of course, Google searches are somewhat different for every person, depending on their Googling history; the way to find out how a website you own is ranked in Google searches on average is to register it with Google webmaster tools (www.google.com/webmasters). But simply Googling your own name is a good way to start—and a good way find the many sites mentioning you that you do not precisely own yourself.

At some point, Facebook or maybe another site might become the main Internet portal for the greater scientific community. But right now, our first encounter with our scientific customers on the Web is generally through Google searches. The list of links that pops up when someone Googles your name—that's the real front door to your scientific business on the Net.

CHAPTER THIRTEEN

The General Public and the Government

Some scientists love to communicate with the general public. It can be fun to see yourself on TV or to hear yourself on the radio. And our magazine articles and television interviews and so on can help remind people about our work and inspire the next generation of science students.

Moreover, if you make a big enough public splash, your story can find its way into the hands of congressional staffers and White House staffers—even the hands of the president. Our elected leaders pay attention to whatever they think might strongly influence their constituents. So if you can make it to a nationally distributed newspaper (like *USA Today* or the *Wall Street Journal*) or to a Hollywood movie, someone on Capitol Hill will probably hear your ideas. These days, even a YouTube video or a post on a major blog like the *Huffington Post* can reach an audience of policymakers.

But never mind all that for the moment. When you generate good press, you also help out the staff at the institutions that fund you. First of all, your good press makes your home institution and funding institutions look good; funding agencies use your press releases to show off their projects. Second, the staff at your funding agencies rely on press releases and popular science articles to provide them with information. They do not have the time or expertise to read the technical literature on everything in their purview. Your press releases can help them catch up on the latest news. Helping out these staff at the institutions that support you is a way of marketing your work to them; in other words, your public outreach is a kind of "inreach" too.

There are some wonderful new how-to books on the topic of how scientists relate to the public, the press, and the government, such as *Escape from the Ivory Tower: Making Your Science Matter*, by Nancy Baron, and *Am I Making Myself Clear: A Scientist's Guide to Talking to the Media*, by Cornelia Dean. So I won't try to cover this topic comprehensively. I just want to share some of the advice I collected and to illustrate how the craft of marketing can help guide you through the media maze. Since influencing the public and influencing government are closely related, I address these two topics in the same chapter. We will talk more in the next chapter about how to market the whole field of science to the public. This chapter will be about marketing your own work in particular.

What Makes Science News?

Before we can get our stories into the news, we first need to sell press officers and science writers on the idea and convince them that people will want to read it. To do that, it helps to understand what the lives of reporters are like, so that we can develop a knack for providing them with the kinds of stories that will make their lives easier.

I was lucky to spend an afternoon at the National Public Radio (NPR) headquarters in downtown Washington, D.C., with science reporter Nell Greenfieldboyce. I asked her what stories she decides to report on. She led me through the NPR newsroom's warren of gray metal bookshelves and showed me her computer screen.

Greenfieldboyce's e-mail inbox was filled with hundreds, maybe thousands of press releases from universities, government labs, companies—everywhere. A few more arrived even during the few moments while I was watching. Bing! Nell looks at each for a few seconds and then generally hits the delete key, unless something catches her eye. She told me that what matters most during those crucial few seconds is the *subject line of the e-mail*. That's what either grabs her or doesn't grab her. For most of the e-mails in Greenfieldboyce's inbox, the subject line was the headline of the press release.

Wow, I thought. If you want to have a successful press release, it had better have a darn good headline. So when I think I have a science result that might be newsworthy, the first thing I do is try to make up a headline for it. Then I try to guess whether the headline is exciting enough to stand out in a busy reporter's inbox.

I've asked many science writers and reporters what they think makes a good headline. Several people told me that if your headline has a superlative in it (like "biggest" or "first"), you're on your way. Also, newsworthy items often deal with issues that hit close to home, or alternately, bizarre, extreme, or science-fiction-worthy topics.

"What stories do people want to read? Stories about instant weight loss and sex."

—Irene Klotz, reporter, *Discovery News*

"Something that's superlative—the biggest or the strongest or the loudest or the most distant. . . ."

—Michael Lemonick, reporter, *Time* magazine, Climate Central

"Whenever we put the word 'Quantum' on the cover, we sell many, many more magazines. Also, we do well when the cover shows something about the origin of humans: anthropology, apes, the missing link. But the word 'Quantum' is like our 'Sex.'"

—George Musser, writer, *Scientific American*

"People want to know how did things come to be the way they are today? How did the universe come into being? How did the Earth form? Is there other life—is there intelligent life out there? People are curious about really weird things that are totally unlike anything we have on Earth. That's why stories about black holes are always popular."

—Robert Naeye, editor in chief, *Sky and Telescope* magazine

"Cute always works. Babies, cats, mammals, charismatic megafauna."

—Nell Greenfieldboyce, reporter, National Public Radio

"If it's something you think you might want to tell your neighbor about."

—Dennis Overbye, science writer, *The New York Times*

As an astronomer, I was already aware of the steady market for the extreme and the bizarre, but I was surprised to learn that, more often, what makes good science news is not the extreme, but the ordinary. For example, Greenfieldboyce showed me two examples of recent NPR stories that were really about connecting science to people's everyday lives: "The Formula for Perfect Parallel Parking" and "What's in That Fish Stick? Give It a DNA Test." She left me wondering if there were some way I could somehow relate my esoteric work on astrophysical disks to, say, the humble pleasures of shopping at Target. Climate Central's senior science writer Michael Lemonick told me, "If it's in the physical sciences it may be hard to argue that the story is going to be of personal importance to people. But then we switch to the awe-and-wonder track."

Of course, the final headline that appears online or in a newspaper or magazine is not generally the one that appears in the press release. Often it isn't even written by the reporter, for that matter. Editors rewrite the headlines reporters turn in.

And editors might not even be writing headlines with readers in mind, not precisely anyway. "Online, it's all about SEO [seach engine optimization] now," *USA Today's* Dan Vergano mentioned in the Marketing for Scientists Facebook group. Wordplay used to be common in science headlines, but search engine optimization has changed that, at least for the time being. Michael Lemonick said, "The more literal a headline is, the more chance it corresponds with something people will be searching for. Irony and puns are rarely things people are searching for."

The Google search engine has a bit of a personality, one that's worth getting to know in general. It prefers pages with lots of original content and with many other pages that link to them, for example, and it has a variety of features that attempt to weed out sites that it considers annoyances, such as those with pages that seem to duplicate other pages. Google is a beautiful thing. It's ironic, though, that one of the main "customers" for our science accomplishments is not a person, but a search engine.

Sound Bites

An article appeared in the *NYTPicker* website last year about a professor who has been quoted in 150 different *New York Times* articles—an amazing achievement.[1] That professor is Robert J. Thompson, the Trustee Professor of Television and Popular Culture at Syracuse University and the founding director of the Bleier Center for Televsion and Popular Culture. When I read the article, I was impressed. I was even more impressed to learn that this wasn't the first article about Thompson's extreme pundit status. In 2007 the Associated Press ran an article about Thompson, quoting the dean of the School of Public Communications at Syracuse about his super media skills. The dean said, "I've seen Bob get 60, 70, 80 media calls in one day."

Sometimes we interact with the press by writing a press release, and sometimes we comment on a story as an outside expert. I called Bob to ask him for advice on serving as an outside expert. The advice he gave me was applicable for both purposes. Of course, he said, being a well-quoted expert is all about being *useful to them*—the press.

The first lesson I learned from Bob is that when the press calls, you need to be cognizant of the fact that the story the reporter is writing is only going to be 500–2,000 words long, and of those words, you are only going to get a few. So to help the reporter make the most out of those few words, you have to learn to come up with sound bites. Most scientists, he said, should have a repertoire of sound bites in their heads.

Now, the term *sound bite* has come to sound derogatory, indicating a kind of oversimplification or dumbing down. But the goal isn't to dumb down. You could just as well think of it as smarting-up—using wit and metaphor to make something obscure into something memorable. As Bob pointed out, *Bartlett's Familiar Quotations* is a book of sound bites. "We have nothing to fear but fear itself." "Russia is a riddle wrapped in a mystery inside an enigma." Even the Gettysburg Address consists essentially of about ten sound bites in a row.

According to Bob, what makes a good sound bite is that it summarizes a lot of information, or a tough concept, very succinctly. Sound bites also often use elements of poetry; as Daniel Pendick, former associate editor at *Astronomy* magazine told me, "A good quote often has a simile or metaphor that helps people understand what's going on." For example, say you were describing an astronomical measurement technique you've used that has a remarkable ability to distinguish two stars that other telescopes could not tell from a single star. Instead of saying, "It lets you resolve two stars separated by a microarcsecond," you might say, "It's like being able to read the date on a penny—if the penny were in California and you were in New York."

Sometimes, science writers talk about a special kind of sound bite called a "pub fact." A pub fact is a quotable fact about your scientific work that somehow appeals to the archetypal everyman. For example:

"The energy of the new subatomic particle is equivalent to that of a baseball traveling a hundred miles an hour." It makes for good conversation around a pint of beer.

In general, paying attention to the way we handle numbers and quantitative information seems to be crucial in sound bites and other communications to the general public and to policymakers. Analogies like the ones above are probably the best way to communicate quantitative information in a sound bite. In *Escape from the Ivory Tower*, Nancy Baron tells us that another effective approach is to use frequencies, like "one out of fifty" or "seven out of a million." In other words, instead of saying "13.7 percent of the participants in the study showed signs of remission," it's better to say "one out of seven."

Sound bites have many uses. Every press release needs a few sound bites. If you find yourself on TV or on radio, or talking to congressional staff or White House staff, you can use sound bites. Sound bites seem to be a kind of natural unit of information in the human brain; if you think hard about it, I bet you'll see that most of what you know about many topics (try the Spanish-American War, or Charles Lindbergh, or popcorn) is essentially a handful of sound bites. Search your feelings, Luke.

In my experience, when I am communicating with the press, and I can't come up with a clever sound bite in a hurry, it helps if I make a comment that reveals some kind of emotion. For example, "When I saw the title of their paper, my heart skipped a beat." I think I must have learned somewhere to suppress these emotions, even in my memory. Maybe that was part of my scientific training, or maybe it was a fear of seeming false. But forgive me if I insist: there is *always* something emotional about your science experience that you can remember, if you allow yourself. If there isn't, you're in the wrong field.

Marketing Science to Congress

The résumé of Colonel Randall Larsen lists Founding Director, Institute for Homeland Security, National Security Advisor to the Center for Biosecurity at the University of Pittsburgh Medical Center, executive director of the Congressional Commission on the Prevention of Weapons of Mass Destruction Proliferation and Terrorism, Expert Witness, 9/11 Commission, author of *Our Own Worst Enemy*, and air force pilot and commander.

But no one of these titles would suffice to sum him up. Larsen is a professional communicator, serving up a steady stream of stories and sound bites to the government, to the press, and to the public on issues related to science and national security.

One rainy afternoon, I sat down with Larsen in his impeccably tidy corner office two blocks from the White House and asked him for advice on how a scientist can influence government. In the throaty voice of a debate-team coach, he walked me through some of the basics.

Larsen said that as a scientist, there are four subcommittees in the U.S. Congress that you need to know. These four subcommittees are the appropriations subcommittee and the authorizing subcommittee for the program that funds you, in both the House and in the Senate. There will be one or two staffers in each of those four subcommittees who are interested in what you are doing—you need to know their names.

Here's how it works. The authorizers are the ones who give the legal green light to do something. When most people think of legislation, they have in mind authorizing legislation. Every government program you can think of—the post office, Social Security, the National Science Foundation, the National Aeronautics and Space Administration, the National Oceanic and Atmospheric Administration, the Department of Energy—was created by some authorizing act. For example, the House Subcommittee on Research and Science Education authorizes funding for the National Science Foundation.

The appropriators live in a different, parallel universe from that of the authorizers. They are the ones who tell the U.S. Treasury to release the funds to do the job. It takes both kinds of congressional committees to make anything happen. Both the House and the Senate have subcommittees called "Subcommittee on Commerce, Justice, Science, and Related Agencies" that handle science funding appropriations.

These two functions sounded redundant to me at first, but here's a different angle on it. The authorizer's job is to dream big. The appropriator's job is to save money. Ideally, they duke it out and try to create a compromise that gets the job done for a reasonable cost. From our point of view, what that means is that scientists have twice as many people in Congress who might possibly be interested in talking to us as we otherwise would—and that's a good thing.

Larsen said that to open the lines of communication, you start by paying attention to the legislation relevant to your subject area that's been passed. Then you use your Google skills and networking skills to track down the e-mail addresses of the staffers on each committee who work on your program (hint: they probably end in "house.gov" or "senate. gov"). Then you send them cordial e-mails.

Larsen suggests you start out by saying something like, "Hey, I bet you were involved in that piece of legislation that just passed: that was a great thing!" That's it. That's the end of the e-mail. That's just to get the conversation going. It's just like what we talked about in chapter 4, on building relationships: you have to start by flirting.

Then, Larsen says, if they write you back, you can offer them your help—i.e., ask permission to move a little closer. "Hey, if you ever have a question about [the subject you're an expert on], don't hesitate to ask." Congressional staffers need expert advice. So if your pet topic keeps coming up in the news or in legislation, you can meet their needs by helping bring them up to speed in your field.

Once you have the ear of a staffer or a congressperson, Larsen said, the best way to move forward with whatever you want to say is to use sound bites. It's just like talking to the media. For example, Larsen told me, "I was testifying in front of the Senate appropriation committee and the impression I wanted to leave is that the Senate is not setting aside enough money to clean up anthrax. So I said, 'Senator Harkin, do you know how much money the Senate has set aside for environmental cleanup after an anthrax attack?' And he said no. So I said, 'Half of what you appropriated for the Marine Corps marching bands.'"

The sound bite is just to pique their interest. If they are interested, they will want to know more. Then you can tell them the details: what's known, what's not known, and what needs to be done. Just be sure, Larsen said, that you offer some way to take immediate action—there is no time to hem and haw. Politicians often feel like they need to take some kind of highly visible action, in order to show their constituents that they are responding to whatever crisis is at hand. "When a senator asks you, 'What should we do?' You say, 'Three things, Senator. And by the way, two of them are good for your state.'"

Now, advising a senator or other policymaker can be a touchy business for us scientists. Though our work may seem to support certain policies or legislation, it always has loose ends and statistical uncertainties, and there are always people in our field who will disagree with it. We may want to take political action based on our data, but we will never really have enough certainty in our findings to fend off all objections or criticism. In *Escape from the Ivory Tower*, Nancy Baron offers a suggestion on how to bridge this gap. The magic words are: "As a scientist, I can tell you that . . . , and as a citizen, I feel strongly that . . . ." If you can speak alternately as a scientist and as a citizen, you can allow for the uncertainties in your scientific conclusions, while still providing a strong voice and a compelling call to action.

Developing Relationships with the Press

You may already have begun to develop relationships with reporters. And of course, marketing is all about building relationships. Science journalist Genevive Bjorn gave me some tips on how we scientists can do this better.

Bjorn told me that an important and often overlooked way for scientists to meet members of the press is to go to science conferences, hang out in the press room, and try to participate in press social events. At science conferences there will often be dinners and luncheons for members of the press; you might be able to join them if you ask the press-room director for permission. She said, "Attending a press social event shows that you are interested in talking to the media. And we find it's a refreshing, low-key way to meet scientists. So go ahead and chat with reporters, tell them what you work on, let them ask you a few questions, and give out your business card. You won't end up getting quoted unless the media knows who you are. And when we need sources for a story, most of us make contact first with scientists we already know."

As you're probably tired of hearing from me, developing relationships involves being thoughtful and genuine, and thinking about the whole person. There's a bit of mind reading to be done to figure out how to do some of a reporter's work for her. Randall Larsen told me, "One thing you have to remember about reporters is they have Little League games to go to. They just want to get their story out the door and go home. So some people wait for the pull; I give them the push." In other words, once you've gotten to know a few reporters, they might appreciate it if you send them news and sound bites before they even ask. "When there's a news story, when you hear about some new thing—I would instantly send text messages to my top three reporter friends and send them a quote about it. That way, they don't even have to call me!"

Besides their thoughtfulness and their proactiveness, part of what makes Randall Larsen and Robert Thompson so useful to the press is their willingness to be quoted experts on a wide range of subjects. To reporters, they offer a kind of one-stop shopping. When I brought this point up with Larsen, he told me that having broad expertise is even more than just a matter of offering convenience to reporters; you have to be an expert on a wide range of other people's work so that you can *promote* other people's work. Too much *self*-promotion looks bad, but working with reporters to promote someone else's science displays a winning generosity of spirit and interest in the field. If you e-mail a reporter four times in a row to help spread the good word about someone else's science, then he or she will be more ready to listen the fifth time, when you bring up your own work.

Becoming this go-to quotable wide-ranging expert requires swallowing a bit of pride, and also putting effort into keeping up-to-date on the news in a broader range of fields than you otherwise might. We scientists can have a tendency to shy away from topics that are not precisely what we have written about in our last papers. And, indeed, the topics we've written about recently are the only topics we are really, really experts in— at the level where we can debate, or surpass, other international experts. That place, the cutting edge, is where we proudly picture ourselves, and where we need to be.

But to be that quotable expert in a broad range of subjects, the level of expertise you need is only the level you would teach to an introductory college class for non-majors. If you've taught such a class, you know that's already more than most people can easily handle. So if you're a college professor, you're already most of the way there—though communicating to the media works better if you talk like you were talking to a friend rather than teaching a student.

Media Training or Improvisational Acting?

One day, someone may want to interview you on television. And television aside, I suspect that upcoming generations of scientists will probably make and post videos on the Internet as a matter of routine, for communication with the public and also with other scientists. Performing in front of a camera seems to have become a valuable, if not essential skill for a scientist to have—and it's becoming ever more important.

There seem to be two main approaches to teaching scientists how to behave in front of a camera, each one recommended by respected experts. Many institutions offer some kind of media-training class for scientists; that's the old-school approach. And now, some institutions are starting to offer improvisational acting classes to scientists, with a similar goal in mind. That's a starkly different philosophy.

I once took a media-training class with about twelve other scientists, led by television producer Geoge Merlis. We sat together all day in a boardroom with our patient teachers, a video camera, and a giant fuzzy microphone. Despite my many hours spent in a studio recording country music, the sight of this microphone made my heart pound.

We learned that, first of all, you are supposed to have an agenda: some kind of idea that you're pushing. That certainly wasn't a stretch for me. We scientists are in the idea business, so I just thought of two or three recent papers I'd written that happened to be on the same topic, and decided that I would try to promote that topic.

Our trainers had each of us write down a page of interesting facts about our pet topic: sound bites. Preparing sound bites is a crucial skill, as I've already mentioned. Then the trainers turned on the camera, interviewed us, and videotaped our responses.

That was useful. I saw how, on camera, it is good to talk with your hands. Using measured hand gestures makes the visual more exciting than a shot of just a stationary talking head, and may even make you better understood and more memorable.[2] While practicing this technique, I also noticed that I had a habit of wiggling my eyebrows when I got nervous or excited, or when I said something I feared might sound self-serving (like the words "Harvard" or "astrophysicist"). On camera, that nervous habit made me look a little bit like the character Animal on *The Muppet Show*.

We were also taught the technique of *pivoting*: staying on-message during an interview, even when the interviewer tries to change the topic. Picture President Clinton on TV, standing before a bank of microphones. A reporter brings up an uncomfortable issue: health care. The president says, "Well, that's a very interesting issue, but I'm here to talk about jobs." He then launches into a sound bite about the economy and never answers the question about health care. As I mentioned in chapter 11, this kind of phrase ("Well, that's interesting, but . . ." or "I can't tell you much about that, but what I can say is . . ." or "Thanks for your comment; you've reminded me of something I wanted to tell you about . . .") is called a pivot.

I came home with a few pages of notes I'd written, full of sound bites. I was ready to pivot and to fire those sound bites at any reporter who came my way. Then, coincidently, I had an interview the following week with a reporter from NPR.

The trouble began when the reporter turned on his recorder, and I whipped out my sound bites. I dodged the first question and launched into a cute factlet about my agenda that I had lovingly prepared for the occasion.[3] But the reporter did not seem pleased by my maneuver. When I recited my canned fact, he visibly winced. But I had another point I wanted to make, so I tore into another prepared sound bite. The reporter started to pull away. I was left questioning myself, my science, and my media training. Pivoting and prepared sound bites may be crucial techniques to use when the stakes are high. But perhaps they are not always the best way for scientists to handle media interviews.

There is another approach. Emmy-nominated television producer Dana Berry once gave me some advice about how to handle a recorded interview. He said, "Don't be afraid to speculate and imagine on camera. There are many cases when I get someone [a scientist] who won't guide me into the possibilities of a certain subject by defaulting to, 'well, we don't know,' or 'I'm not the expert on that subject.' So I'm forced to ask the question in different ways, and sometimes I still can't get past the facade. A TV interview is not a peer-reviewed paper, so my advice is: Don't be afraid to open up. Dream, speculate, and share your excitement . . . in other words, be a human."

I've tried to follow Berry's advice. I took some acting classes when I was in high school, and I've found it useful to draw on that experience.

Learning acting was also important to physicist and Nobel laureate John Mather. He told me, "Another piece of advice I would give is study acting. There are some great little books about acting that you can pick up." He recommended *Acting: The First Six Lessons,* by Richard Boleslavsky.[4]

Lately, a new kind of training class for scientists has appeared on the scene, one that Berry and Mather would approve of. Actor Alan Alda has teamed up with faculty members at Stony Brook University to offer acting classes to scientists. Alda told a reporter that his classes are a way of helping scientists connect with the general public through emotions. "Emotion is so important," he told a reporter. "In scientific communication, emotion is probably trained out of us, but there's no reason why it can't be included."[5]

Alan Alda's acting classes for scientists emphasize a particular kind of acting: improvisational ("improv") acting. In improv acting, there is no script. Picture yourself onstage, in character. When another actor walks onto the stage for the first time, you don't know quite what his role is going to be. But nonetheless, you have to make the moment seamless, fluid, and natural—for both your own character and the new character.

To achieve this seamlessness, a good improv actor accepts whatever new information about the plot and situation emerges when the new character opens his mouth. If the new character asks you onstage if his dinner is finally ready, it doesn't matter if you don't feel like acting the role of a cook; you are one now. A bad improv actor is one who tries to control the moment, and force the scene into a preconceived mold—like a politician being asked uncomfortable questions by the press and trying to stay on-message.

Clearly, both kinds of training—media training and improv acting—can help people become better communicators. But as Randy Olson has pointed out, they seem to be at odds. A media-training class teaches you to lead with your mind. An improv acting class teaches you to lead with your heart. A media-training class suits the ruler archetype. An improv acting class helps you tap into the everyman. Which one is right?

I think that what marketing teaches us is that neither extreme is right or wrong. President Clinton was known for his amazing ability to connect with people—yet he wanted to look like a ruler, and to separate himself from the issues his opponents were using to brand themselves. Scientists, on the other hand, are often stereotyped as distant and aloof. Alan Alda's class is about learning to be spontaneous and open to the moment—qualities that help scientists appear more approachable. What matters is choosing the path that suits your background, your goals, and your audience.

It seems to me that the ideal media preparation for scientists might contain some of each element: both kinds of classes. That preparation would help us with *positioning* ourselves—recognizing and countering preconceptions—in front of a range of different audiences. I find that when I am in a situation where I'm probably being perceived as being not in command of my subject, the media training approach serves me well. For example, if I am answering questions at a conference of aggressive colleagues who are unfamiliar with my work, I may need to pivot away from their grandstanding questions. But when scientists appear on radio and television, we face very different kinds of preconceptions: stereotypes that we are aloof, insensitive, and full of ourselves. (We'll talk more about these stereotypes in the next chapter.) We can better work against those preconceptions of scientists by using what can be learned from improvisational acting.

If you have taken a media-training class, you have probably prepared a handy list of talking points about your work. Maybe we scientists should also be taught to prepare a list of *feeling* points. What are our hopes and dreams? What do we love about our work? What do we want to see accomplished before we die? Of course, it wouldn't work to read these points from a sheet of paper during an interview. It wouldn't even work for us to memorize them; they would sound canned. For the best effect, we will still have to improvise.

Writing a Book

As I have been interviewing experts for this book, several people have told me how a book can be a powerful science marketing tool. If the book is any good, it establishes you as an expert in the book's subject—a publicly credible expert worth calling for quotes and advice. Cristina Eisenberg told me how writing a book, *The Wolf's Tooth*, directly helped her raise funds for her science.[6] She said, "I just spent some time with the chairman of Chevron, and he had read my book from cover to cover. He wanted me to show him my field sites on the ranch and offered to support some of my research. If I hadn't written that book, its unlikely that that person would have responded the way he did!"

As I mentioned earlier, science blogs are no longer the novelty they once were. But I can tell you from experience that writing a blog is one good way to start getting into the rhythm of writing a book. A blog lets you get feedback on your material and simultaneously helps you build relationships with people who care about your subject. That's more or less how I wrote this book, publishing my ideas—and questions for my peers—on a blog inside a Facebook group.

Starting your book project by writing a blog has a few other practical advantages. It keeps you on a schedule because you'll feel like you have to post something new regularly. Writing a book is a big, risky project that may never lead to satisfaction; you'll be tempted to give up halfway. But writing a single blog post is a small project, and one that can be satisfying all by itself.

CHAPTER FOURTEEN

How to Market Science Itself

 Many of us scientists are concerned about the state of science as a discipline. We fear the failures of science education and the spread of creationism. We are exhausted by politicians who deny that humans have caused global warming, or who talk about "mice with fully functioning human brains."[1]

The state of public appreciation for science seems particularly gloomy in the United States. As Colonel Randall Larsen said to me, "In this country, I'm very frustrated that people take their advice about vaccines from Rush Limbaugh, Reverend Al Sharpton, Dierdre Imus, and Jenny McCarthy, former playmate of the year." These attitudes are matched by an apparent decline in American competitiveness in science and technology. *New York Times* science writer Dennis Overbye told me, "As far as the world is concerned, Europe has risen from the ashes and it's pretty much equal to us now, and Asia is coming on strong. Overall, I think there's more research going on, more doctors being trained. But most of them won't be in the U.S. anymore."

Here are more blood-curdling sound bites about science knowledge, education and investment in the United States from a 2010 National Academies report called *Rising Above the Gathering Storm, Revisited*.[2] The report starts with five pages of grisly factlets like these:

- Forty-nine percent of U.S. adults don't know how long it takes the Earth to revolve around the sun.
- More U.S. college grads take home degrees in the visual and performing arts than in engineering.

- The federal government's annual research investment in engineering, math, and the physical sciences "is now equal to the increase in United States health-care costs every nine weeks."

Many times, hearing statistics like these, I have wished I could just grab people and shake them—and feed them a giant science sandwich. It's an emotional subject, and it almost makes me feel violent.

But there are hundreds of thousands of scientists in the United States, and we're supposed to be smart. Instead of just feeling upset, frightened, or paralyzed by these developments, perhaps we can use our powers to craft a sales and marketing plan to help win converts to our cause. That's what this chapter is about.

Our Audience: Mythbusting

To devise any marketing plan, we must begin by considering who our audience might be. So let us start along this path by thinking broadly about the non-scientists who live in the United States. What are their needs? Science: what's in it for them?

With roughly half a million scientists (excluding computer scientists) in the United States, there are roughly six hundred non-scientists for each us in the entire rest of the country.[3] That's our audience, at first glance. That's not so big; you may already have 600 Facebook friends.

Given how small this number is, maybe we scientists could rescue science simply by stepping out of our social and professional cocoons and carrying our message to the public in person. And maybe we each don't have to meet all 600 people in our allotment ourselves if we have a message that spreads on its own. Studies show that when an idea goes viral, the inflection point in the adoption curve occurs at somewhere between a 5 percent and 20 percent level of adoption.[4] So maybe that means we each need to get to know only 120 non-scientists to plant the seeds that will generate the support we need.

Of course, scientists, like all people, tend to socialize mostly with other like-minded folks. We are often content to avoid things like churches, small towns, and rural areas. That leaves many people in the world in the sorry state of never having met a scientist personally. But that's something we can change.

So let's take a closer look at these 120 non-scientists we each need to get to know. It's a diverse crowd, to say the least. Of these 120 people, roughly 80 will not have gone to college. Forty-eight will hold jobs in industrial or manual labor. Fourteen will live below the poverty line. Twenty-four will be under the age of fourteen. Yes, many of us scientists consider ourselves fluent in pop culture and the glorious range of different American lifestyles, heartaches, and dreams. But really—how well do you understand the needs and desires of all these various groups? How well could you craft a sales pitch for Science that would answer their inevitable (if silent) question: "Science—What's In It For Me?"

I do know one institution that delivers this pitch flawlessly, vividly demonstrating the power of scientific inquiry in a way everybody can understand: the show *Mythbusters* on the Discovery Channel. On each *Mythbusters* show, hosts Adam Savage and Jamie Hyneman choose a myth or a rumor, and test it with an experiment. Each show demonstrates the scientific method; the myth is the hypothesis, and the test is designed so that it will be capable of falsifying the hypothesis.

The first thing I like about *Mythbusters* is that it offers something for everyone: the myths they put to the test are familiar and believable. For example, we've all heard that a penny dropped from the Empire State Building can land with enough force to kill someone (Episode Four). I'm pretty sure I believed that idea in third grade, and then maybe I still believed it as a PhD, lacking relevant data on which to base a deeper analysis. I might have even heard it first from a teacher. On *Mythbusters*, you get to find out if your favorite myth is true or not; that's a great big WIIFM.

The second thing I like about *Mythbusters* is the storytelling—in particular, the element of suspense. Only at the end of the show is the myth pronounced Confirmed, Plausible, or Busted, depending on the outcome of the test. The hosts act as though they do not know what the results of the test are going to be; they are generally dazzled by the outcomes of their experiments.

It might be possible to make some progress on the problem of public image with some combination of sustained personal contact and *Mythbusters*-inspired educational materials. The "1000 Scientists in 1000 Days" initiative being launched by *Scientific American* takes this approach by arranging for scientists to visit schools.[5] We will discuss this option and its pitfalls more in a bit.

The penny-drop myth, by the way, was busted. When the Mythbusters fired a penny at terminal velocity into a ballistics gel head with a human skull, they were unable to penetrate the skull. Then, when they visited the Empire State Building, they realized that updrafts and the roofs of lower floors would probably prevent a penny dropped from the roof from even reaching the ground.

Positioning Ourselves

To take the next step in our marketing plan, we must think about the preconceptions our audience has about science and scientists, how we can recognize them and help people move past them, if necessary. In other words, we must think about our *positioning* (chapters 3 and 11).

The public's preconceptions of scientists have been widely discussed lately, with good reason. You don't have to watch too many classic movies to find out that scientists are people who dress badly, have big egos, and speak in a language that nobody understands, including other scientists. Sometimes, scientists are even mad and out to destroy the world. Another way to put it might be to say that to non-scientists, scientists sometimes take on the shadow side of the sage archetype—bad advice-givers who spout only confusion.

> "Scientists are often portrayed according to one of two stereotypes. You're either the mad, evil genius, meddling with forbidden, inner secrets of nature, or, you're a loveable but fumbling Mr. Magoo, so preoccupied with your research you don't have time or concern for basics like grooming. These are the sort of impressions you're dealing with."
> —Dana Berry, Emmy Award–nominated television producer

> "The characters in the TV show *Big Bang Theory*, especially that Sheldon character. That's how the public tends to view scientists."
> —Colonel Randall Larsen, science communicator
> and policy expert

We can combat these negative images in a few different ways. One approach is to recruit or hire likeable spokespeople. For example, Randy

Olson hired actor/comedian Jack Black to star in his public service announcement on the decline of the Earth's oceans; Black's image on the cover of the video helped score Olson millions of dollars of free television airtime. We can also try to change the way we project our images by, for example, taking acting classes and by showing our more spontaneous sides, as we discussed in chapter 13.

But before we focus too much energy on this specific problem, note that the public image of scientists is far from hopeless; many Americans still harbor a deep respect for scientists. An often-cited 2009 poll of U.S. adults measured the prestige of scientists compared to twenty-one other professions.[6] The study found that 57 percent of U.S. adults consider scientists to have "very great prestige." According to this poll, scientists have nearly the most prestigious occupation in America, second only to firefighters.

Also, the public image of scientists is changing. Dana Berry told me, "I think the image of scientists is in flux. You watch science documentaries nowadays and see attractive young people brimming with enthusiasm and expertise." I think many scientific institutions have begun tapping spunkier—and younger—scientists and interns to pose on camera. Jennifer Ouellette, former director of the Science and Entertainment Exchange (www.scienceandentertainmentexchange.org), told me that Hollywood is keeping up. "While you still see some of the standard tropes— the mad scientist, the nerd in glasses—thanks to *CSI* [The CBS television series *Crime Scene Investigation*], there are all kinds of scientists on TV now. They are male, female, white, black, Asian."

Though the public image of scientists may sometimes be tricky to pin down, it's important to remember that we should *always* think about positioning ourselves, even when we are lucky to have an audience that is predisposed to like us. Positioning is about allowing and appreciating whatever feelings people may have about you and the issues you're raising, good or bad. So it can be endearing to say once in a while, "Hey, I know I'm kind of a nerd," if you suspect you're up against some negative stereotypes. Or if you can surmise that people's conceptions of scientists in your field are likely positive from watching *CSI*, it might be time to say something like "I'm just like the guys on *CSI*." Then, to be humble, you might add the words ". . . but uglier."

Robert Thompson, the founding director of the Bleier Center for Television and Popular Culture at Syracuse University, gave me another straightforward example of how a scientist can position himself—in this case when he is facing criticism. "Let's say you're being called to the carpet on the Glenn Beck show because the government is giving all this money to your project and people don't understand what the project is about, so they think it's a waste. You have start by saying, 'Well, if I heard my government were giving all this money to something that sounds so esoteric, that would sound ludicrous to me too.' Once you've connected with people in this manner, you can move ahead with your explanation of why funding your project is anything but ludicrous." Thompson added, "If you say in your speech the words 'I sometimes struggle with this as well,' you'd be amazed how that can ingratiate you with an audience."

A Good Brand for Scientists: Purple Cow

When Joshua Wurman was a postdoc at the National Center for Atmospheric Research (NCAR), he had a crazy idea. Instead of just studying hurricanes and tornadoes from afar, why not put Doppler radars on trucks and drive them into the storms?

Wurman tried to convince the people in his lab to do it. But they were completely uninterested. So he talked to the senior people in his field to see if he could win support for the idea. They weren't interested either. Wurman told me, "The senior scientists, engineers, technicians, etc. looked at us and told us the truck would tip over and kill us and that we needed to do wind tunnel testing, etc. etc. There was no funding for us to do wind-tunnel testing or anything like this! And, I was impatient."

Wurman accepted a faculty position at the University of Oklahoma and left the lab at NCAR, taking with him a U-Haul truck full of leftover radar parts. He and his wife and a few graduate students stayed up all night lashing them together. Somehow, their creation worked. The new invention, the Doppler-on-Wheels, became an SRI for Wurman. He said, "The prototype was lashed together literally with duct tape. But it revolutionized the field. We were able to see what was inside a tornado, for the first time make 3-D maps of tornado winds, watch how they evolve, show that this was possible—not just a postdoc's overambitious fantasy."

As you may remember from chapter 5, about branding, I am all in favor of postdocs' ambitious fantasies. Joshua Wurman, like Cristina Eisenberg, is an example of what marketing guru Seth Godin would call a Purple Cow. So is Wurman's Doppler-on-Wheels. Wurman, Eisenberg, and Doppler-on-Wheels are brands so remarkable that they are newsworthy. Indeed, Josh's Purple Cow story may sound familiar to you. That's because Wurman, and the Doppler-on-Wheels, are stars of Discovery Channel's reality series *Storm Chasers*.

On *Storm Chasers*, Wurman and a team of rugged but slightly scared-looking guys pack up specially armored, futuristic vehicles and drive all across the American Midwest in search of tornadoes. Wurman, sometimes aided by some animations and a voiceover, briefly describes some of science of extreme weather. There are always ominous scenes of overwhelming natural force, of violence and destruction.

But while *Storm Chasers* is partly about storms, it's mostly about passion. What else would drive someone to venture right into a deadly tornado? The show's website talks about people who are "hooked on the adrenaline rush." One episode was called "A Passion for Extreme Weather." That passion is the hallmark of the Purple Cow.

Now, sometimes the show reaches too far in its quest for the human element. Wurman told me, "Anyone familiar with actual storm research and even storm chasing can see that there are story lines which happen on the show which are not real. . . . What they really want is the Real Housewives of Tornado Alley. They have found amateur storm chasers who are willing to do the acting and exaggerating for them in these types of shows." Wurman explained to me that it's always like this with television shows, especially long-running reality shows like *The Deadliest Catch* and *Storm Chasers*.

But ultimately, I think that this kind of show is good for science: it helps brand scientists as brave leaders with extraordinary passion, as Purple Cows who look good on television. Wurman's story is the modern-day equivalent of the irresistible tale of how Benjamin Franklin risked his life for science, flying a kite in a thunderstorm. I think that this image is good for science as a whole. We can continue to foster this kind of brand for science by continuing to live it—by being daring, dedicated, and extreme.

Archetypes and the War on Science

Let us continue our discussion of branding by examining the archetypes involved in the national conversations about scientific issues. Given the media's hunger for controversy and drama, we shouldn't be surprised that science is often described as being engaged in a war—with politicians, with religion, and so on. For example, in *The Republican War on Science* and *Unscientific America*, author Chris Mooney paints a picture of a country torn by an ideological war with science on one side and conservative politicians on the other. In his public talks, Mooney calls for a new kind of "ninja" fighting for science, one who can outflank conservative soldiers like Marc Morano, the fast-talking climate-change critic.

The events and statistics that Mooney cites are alarming; it's easy to see why he is compelled to describe the situation as a war. It seems to me that Mooney himself is one of the ninjas he calls for, fighting for the triumph of experimental verification over dogmatic belief. We scientists owe him our allegiance—and it may be tempting for us to adopt his style as well, becoming archetypal warriors marching into battle.

But often to engage in a public "war" where someone else dictates the terms of the fight is ultimately to lose the battle. For example, if you are attacked in a debate on television, and you simply argue back, I claim that you can never win more than about one third of your audience. Maybe one third will take your side, one third will take the other person's side, and one third will remain undecided. That's partly because some people always root for the underdog. And it's partly because the reporters who edit and comment on your interview will want to shape the story in such a manner that it divides the audience in roughly this way—in the interest of entertainment and the flawed ideal of "balance."

Alan Leshner, the Chief Executive Officer, American Association for the Advancement of Science, has also argued that a warrior stance has limited utility for scientists. In a lecture at a meeting on the role of the life sciences in preserving national security, he said, "Only scientists are stuck with what science shows. . . . Never debate an ideologue. You are not allowed to make things up. They are allowed to say whatever they want."[7]

Fortunately, the theory of archetypes (chapter 6) gives us a different view of the same scene. A marketer is always competing, but is only sometimes at war. There are many strategies for competing in the marketplace of ideas.

Let's look at how an analogous situation is handled in industry. Pepsi, the second-ranked cola company, has a history of picking fights with Coke, such as the Pepsi Challenge ad campaign of the 1980s. But Coke doesn't argue with Pepsi. Coke is the brand leader. To enter into an argument with Pepsi would be to stoop to Pepsi's level and shatter Coke's innocent archetype. Instead, Coke's ad message has been that "Coke is it" and "Coke, it's real thing."

This model seems to apply to scientists who are under attack in the media. In the business of selling factual arguments, scientists are the brand leaders; every school child knows that science history contains a long list of mind-stretching, logical arguments. The concept of "scientific fact," meaning a repeatedly confirmed observation, does not mean precisely the same thing as the single word *fact*, but it's close enough that most people consider the two concepts synonymous, or perhaps consider a scientific fact to be a "better" kind of fact. You might even say that we scientists "invented" facts, or at least a certain kind of fact, just as Coke more or less invented the cola-flavored soda. This brand leadership, combined with the prestige of scientists as indicated in the Harris poll, suggests that scientists should have the confidence never to argue over facts with our critics on television. We don't have to argue in public because we are the real thing. We can use whatever airtime we have to tell stories about our work instead; storytelling is a more powerful medium anyway.

Besides following the model of Coke and Pepsi, we have other archetypal options to choose from, and we can pick whichever one suits us most naturally. For example, Chris Mooney and climate-change denier Mark Morano are both heroes/warriors in public. So are many scientists. That leaves eleven other archetypes that we can use to help us connect with whatever audience may be paying attention.

Maybe the scientists involved in the climate-change debate (it feels like that's most of us, at some level) could evoke one of these other archetypes, something very different than the warrior. To help evoke the caregiver, for example, we could speak in terms of protecting the next generation: when it comes to taking care of your children, you don't take any chances. We could pose for pictures with polar bear cubs, as World

Wildlife Federation scientist Geoff York has done.[8] Ideally, some of us might start charities to help people or animals suffering from hurricane damage or from diseases made widespread by climate change.

In the "balanced" world of the media, where every voice has equal airtime, the more different, distinct brands a community puts forth, the more exposure it will get. This notion applies to the brands of spokespeople as well. Not every media outlet would define "balanced" the same way, of course, and may not feel compelled to recognize all the brands. But overall, the more brands and archetypes we can bring to bear, the more exposure we will get. For example, it would be nice to have a caregiver spokesperson, an innocent spokesperson, a ruler spokesperson, and so on, all speaking about climate change at once. And whatever new implication of climate change you find emerges from your research, please name it, illustrate it, and share it. The more real issues related to climate change we can raise, market, and brand, the bigger the portion of the American psyche we will occupy.

Our Marketing Funnel: The Neglected Middle

We've talked about our audience, how to pitch science to them, and how to brand scientists. But the heart of any marketing campaign must be the marketing funnel. To market any product, you must go beyond making a good first impression; you must draw them through your funnel, making them more and more invested in your efforts. Likewise, in order to market science, we must draw people in, from layperson to collaborator to advocate. Only by reaching out to customers at every stage of the funnel so that you steadily generate new advocates can an idea go viral or sustain people's long-term interest. And here's where our real trouble lies.

We scientists are good at appealing to other scientists. We're also generally aware of the need to reach out to the layperson. But it seems like our community often gives little heed to drawing people closer in once we've dazzled them with an animated television show or flashy headline. Science writer Michael Lemonick, who has written for *Time*, *Discover*, *Scientific American*, and Climate Central, told me, "I don't want to knock people's writing about science, but if I think I'm raising the level of science literacy in America by doing it [writing about science], I think I'm kidding myself badly." He argued that even the best television shows about science also have limited power. "People talk about how wonderful it was when Carl Sagan was out there. They say people were talking about science back then! I think that's a crock. I guarantee you: the people who were alive at the time, if you ask them about Carl Sagan, they will remember only four things. He had a show named *Cosmos*. He went on the *Tonight* show a lot. He had a geeky voice. And he went around saying 'billions and billions.'" Even Sagan had limited power to help Americans understand and take ownership of science.

The most sinister example of the neglected middle of our marketing funnel is the failure of Science, Technology, and Mathematics (STEM) education in America. Children in the United States have been learning their math and science from teachers who aren't trained in these fields. And it shows. For example, the World Economic Forum Global Competitiveness Report says that the U.S. ranks 34th in quality of primary education among developed countries, right behind Costa Rica.

This trouble has created a population of voters who don't understand science. Some lucky students who make it to college may be rescued by the high-quality science classes offered in U.S. universities. But universities have not sucessfully reversed the damage done to the majority of U.S. children—and many of these children are now adults. As *Unscientific America* says, 46 percent of Americans deny evolution and think that the Earth is less than 10,000 years old.

Maybe Americans respect scientists, but they don't understand science, and they don't participate in science, so they don't advocate for science. I would even say that our fellow citizens feel left out. The hole in our marketing funnel is a kind of hole in our nation's heart, a wound we've arrogantly helped inflict. Tea Party leader and presidential hopeful Michelle Bachmann hinted at the pain of this wound in comments quoted by the *Stillwater Gazette* in 2003: "I'm not a deep thinker on all of this. I wish I was. I wish I was more knowledgeable, but I'm not a scientist."[9]

Ultimately, the way we can remedy this problem is to change our way of thinking. I mentioned above that every scientist in America needs to reach out to about 120 non-scientists to turn the tide of public sentiment. I used to think that a good use of my public outreach time would be to go out and give a public talk to 120 people in one shot and call it a day. Now I understand that, somehow, I need to find a way to get to know my allotment of non-scientists in some kind of personal way—ideally a collaborative way. I'll have to stay in touch with people, listen to them, work with them, and help them get a little closer to the point where they can give their own talk about science. This is a much bigger task than just visiting a school once or giving a public talk. But if we scientists don't follow through like this, we aren't marketing. We're just self-promoting.

Crowdsourcing and Citizen Science

So we can write all the magazine articles we want, make all the videos we want, show up on television 24 hours a day, and we will still only be working at the mouth of Science's marketing funnel. But what if there were a way to collaborate directly with the public, to involve large numbers of people in actually doing science? Such a massive collaborative effort might carry all these non-scientists right through to our funnel's stem. Wonderfully, some scientists have figured out how to do just that.

In 1999 Dan Wertheimer at the University of California–Berkeley launched SETI@home, a program that allows anyone with a home computer and access to the Internet to help the SETI project search for extraterrestrial intelligence. SETI, the Search for Extra-Terrestrial Intelligence, scans the sky at radio wavelengths, looking for transmissions that could be signs of intelligent life elsewhere in the galaxy. Users download free software that runs when their home computers are otherwise unused, like a screen saver does. The software downloads chunks of radio telescope data, 107 seconds long, from a database at Berkeley, and runs Fourier transforms on them to look for unusual signals. Then it sends the results back to Berkeley to be integrated. Since Wertheimer's program started, over five million volunteers in over two hundred countries have together donated over three million years of computer time to the project.

Since then, David Anderson and others have developed the Berkeley Open Infrastructure for Network Computing (BOINC), a framework that allows scientists working on other projects to interact with the public in this innovative way.[10] Thirty-eight different projects have signed up, including Folding@home, Genome@home, Evolution@home, and FightAIDS@home, and the results have led so far to more than eighty publications. Not every scientific computing problem can be easily divided up so that each volunteer's home computer can work on it separately. But the technique lends itself well to problems that are essentially some kind of search, like searching for signals from ETI (extra terrestrial intelligence) or searching for prime numbers. The BOINC project represents a spectacular example of relationship building, making connections between millions of people who now have something in common that they didn't have before; they all help out scientists by using their home computers. And those millions of people are now invested in scientific progress in a way they weren't before. That's good marketing.

It gets better. Donating computer time to science is exciting to some people, but it has a limited appeal; for example, one survey of SETI@ home users showed that they were 92 percent male. Some new science projects appeal to a broader spectrum of people and let them contribute their own thoughts to it, not just computer time. Some projects virtually allow laypeople to briefly become scientists themselves.

Kevin Schawinski and Chris Lintott were working with a large survey of galaxies from the Sloan Digital Sky Survey, trying to classify them based on their shapes: elliptical, spiral, irregular, and so on. They ran into a problem. The computer software for making these classifications kept making mistakes. But, sitting in an English pub in 2007, they hatched an idea for an interactive website that would solve this problem and also engage vast numbers of volunteers to perform, in a way, as actual scientists themselves.

That July, Schawinski and Chris Lintott launched a website called GalaxyZoo.org (Figure 14-1). *Galaxy Zoo* shows images of galaxies to whomever surfs by, and asks her to classify them based on their shapes. The human brain, it turns out, is better at sorting out shapes than even the best computer algorithms. And after twenty different people classify the same galaxy, the results are much more accurate and consistent than any automatic computer classification system can provide.

To use SETI@home, your first step is to "read our rules and policies." By comparison, using GalaxyZoo.org is a walk in the park. *Galaxy Zoo* invites people to start looking at pictures of galaxies right away, with no red tape to slow them down.

The exercise of classifying galaxies itself is not unpleasant—maybe like playing a game of solitaire or minesweeper. And it carries the allure of the undiscovered. As Lintott said, "These images were taken by a robotic telescope and processed automatically, so the odds are that when you log on, that first galaxy you see will be one that no human has seen

*Figure 14-1   A screen shot from the Galaxy Zoo website (Galaxyzoo.org),
inviting users to help classify images of galaxies from the Hubble Space Telescope.*

before." When a new galaxy appears on the screen for you to classify, you
feel momentarily shocked to be tasked with such a monumental decision,
as though you were sealing the fates of the galaxy's trillions of inhabitants
with the click of your mouse.

Within 24 hours, the new website was receiving 70,000 classifications
per hour. Within a year, *Galaxy Zoo* had received more than 50 million
classifications from over 150,000 people. Will all that help, Schawinski
and Lintott finished their project much sooner than they anticipated, and
they soon set up successor projects to classify images of galaxies from the
Hubble Space Telescope.

American Idol for Scientists

But that's not the best part of *Galaxy Zoo*'s science marketing story. In
2008, a 25-year-old Dutch schoolteacher named Hanny van Arkel was using the website and came across an object that looked nothing like the

other galaxies: a group of deep-blue glowing blobs.[11] When she failed to match the object to any galaxy types described in the *Galaxy Zoo* classification tutorial, she e-mailed the site's webmaster.

Lintott and colleagues published the discovery, referring to the strange image as Hanny's Voorwerp (*Voorwerp* is Dutch for "object"). Then they and others began making more observations of the object, now including Van Arkel in the resulting publications. The object's origins are not crystal clear, but a leading explanation is that the object is a gas cloud heated by the jet of energetic particles from a nearby black hole.

Van Arkel has now traveled all the way through the marketing funnel of Science, from stranger to collaborator to advocate. She began as a music buff, knowing next to nothing about galaxies. She read a book about astronomy primarily because it was written by Brian May, the astrophysicist who also happens to be the former guitarist of the rock band Queen. That step led her to *Galaxy Zoo*, to collaborations with Lintott and colleagues, and to press release after press release. Now she has her own blog, www.hannysvoorwerp.com, where she writes about her experiences going to astronomy meetings, planetariums, meetings of the American Astronomical Society, and working with astronomers to interpret new images of her eponymous object.

I can't help but think of how Van Arkel's journey parallels the journey of Cinderella, or the unknown singers that become big stars on the television show *American Idol*. At some point, a story like that can itself become an attraction, drawing even more people to learn about the underlying event. It's irresistible—the story of a dream coming true. What kind of heartless politician could cancel a project that generates stories like that?

Galaxy Zoo's founders describe their project as "Citizen Science" to indicate that the participants are actually doing science, themselves. Another term for the same phenomenon, more popular in the business world, is *crowdsourcing*. Wikipedia unglamorously defines crowdsourcing as "outsourcing tasks, traditionally performed by an employee or contractor, to an undefined, large group of people or community (a crowd), through an open call."

Whatever you call it, I think this mode of interacting with the public illustrates science marketing at its best. It portrays the job of the scientist as a job people would dream of one day having—Van Arkel's dream come true. The site Scienceforcitizens.net now lists almost 300 searchable citizen-science projects. Maybe someday most science projects will attempt to incorporate a citizen-science component. I think that could be a glorious day for science and science education.

A Seamless Flow

After trying to create a marketing plan to revive U.S. science, the main conclusion I have come to is that we need to focus on the neglected middle of our marketing funnel—the students and adults who are aware of science but never became active participants in it, or in the national conversation about it. Now, some methods of pursuing this neglected middle are already well understood in the science-education community and elsewhere. It seems widely recognized that we need Science, Technology, Engineering, and Mathematics (STEM) education reform, and more scientists to visit classrooms. Citizen science programs like *Galaxy Zoo* and Fold.it that include nonspecialists in the creation of new scientific knowledge are catching on.

But what I think hasn't yet been appreciated is that our marketing funnel must *flow*; we need to keep drawing in people at all stages of involvement. For example, a company that sells snack chips tries to engineer a flow from advertisement to first purchase to addiction and advocacy. It writes ads to make you familiar with the product, it builds in-store displays to get you to buy, and then it makes the chips crunchy so you enjoy them and get everyone around you snacking. Likewise we must create a seamless flow from gee-whiz science news to well-educated K–12 students, to citizen scientists and science fans and professional scientists who keep spreading the good word. That probably means blending citizen science, public education, and outreach such that each can motivate the other. A simple example of how to generate this flow might be putting a link to a citizen-science project or relevant blog at the bottom of every press release.

The classroom is a crucial place where we must improve the flow throuh our funnel. In the classroom, many students encounter science for the first time, but then get the message that science is not for them. In fact, there's no reason for anyone to feel that way in this age of the Internet. We can encourage students and teachers to log in to the citizen-science websites together, classifying galaxies and folding simulated proteins side by side. We can encourage students and teachers to comment meaningfully on science blogs—together. We can award scholarships to top young citizen-scientists. After we scientists visit a classroom, we can continue to interact with the students online.

Clearly, creating all these new interactions will take plenty of time and energy on a national scale. As a scientist you might sometimes feel that the situation is hopeless and that there is no clear way forward toward achieving these goals. But there *is* a way forward. As usual, whatever we want to market, we can start by having a conversation.

CHAPTER FIFTEEN

Starting a Movement

Pensri Ho arrived at the University of Virginia (UVA) as a new assistant professor with a joint appointment in the departments of Anthropology and American Studies. Her very first week, two students appeared in her office to ask if she might be interested in helping establish a new Asian American Studies Program at the university. At first Pensri said no: she couldn't afford to invest time in a project like that before she got tenure. But then after an hour chatting with the eager students, she changed her mind and decided to take up the cause.

To build support for the program, Pensri decided to use a technique from her anthropology training called "snowball interviewing." First she talked to a few colleagues and students who she knew were likely to be interested in helping establish the program. She asked them what courses they thought it should include, and what roles they might like to have in it. At the end of each interview, she asked for names of more people she could talk to.

After a while, Pensri had accumulated a long list of names and interviews. Some people she talked to were merely polite listeners. But a few faculty members actually offered to teach classes in the new program, and two deans offered to help find money for the program. Pensri pressed on, sometimes returning to those key people who had offered help to get more input and to keep them informed of her progress.

At first Pensri was discouraged by the odds stacked against her, the noncommittal responses, and the many administrative obstacles. She wondered if this cause was worth jeopardizing her tenure prospects for.

Then, one day, she found herself overhearing other people talk about her interviews—and about the program. She had generated something marketers call "buzz." At the same time, she realized that she had found enough core supporters to provide the manpower and tools the program needed in order to fly once it had the green light; she had both a consensus about what to do and the volunteers to do the work.

Pensri set her hesitations aside and kept interviewing and growing her snowball. Then one of her contacts introduced her to the university's chief officer for diversity and equity. A dean introduced her to the university's chief financial officer. Together, these people had the power to step over the administrative obstacles and make her vision a reality.

Pensri left UVA for personal reasons one year later. But by the time she left, UVA had a new Asian American Studies Program, including a new senior faculty position and an undergraduate minor. Meanwhile, Pensri had earned a reputation for jumpstarting and leading new university programs; that was a powerful brand that helped her get her next job.

In this book, we have looked at many of the basic concepts of marketing, and we have looked at many specific applications of these ideas to the business of science. But we haven't quite put everything together yet. The story of how anthropologist Pensri Ho helped launch the Asian American Studies Program at the University of Virginia is one example of how to use marketing to start a movement. That's what I want to talk about now.

The Product Development Cycle

Here is one recipe for how to put together all the tools of marketing. It goes by various names, like the Product Development Cycle (PDC) or the Product Life Cycle (PLC).[1] It's the procedure that a company like Kraft might use to develop and market a new snack food or Microsoft might use to develop and market new software. It's the story of TCHO, the chocolate company we discussed in chapter 4. And it seems to me that this script isn't just about selling potato chips, chocolate, or even spreadsheets. It's also, more or less, the script for Pensri Ho's story above, and the basic blueprint for the marketing scientist.

There are six steps in the Product Development Cycle:

1. Learn the trade.

2. Come up with an idea for a product and a brand. The product might be a new flavor of snack chips, a proposal, a research question or research tool, a new department, or it might simply be you.

3. Find a small group of people who might be interested in the product. Use free samples and props to start conversations with them and sell your idea. Get feedback on your idea and refine the product.

4. Widen the network of people who are interested in the products; move the people already in your network down your marketing funnel. Continue to get feedback from your network and use the feedback to refine and develop the product.

5. While you have a group of repeat customers, ask for resources—investments, opportunities, or just payment—in exchange for your product. Use these resources to expand your business. Encourage your community of customers to interact with each other.

6. Repeat steps 4–5 until the market is saturated, your product is obsolete, you get bored, or you think of a better idea.

Note that this algorithm has two processes running in parallel. You develop your product and at the same time you develop relationships with your customers. If you are meant to win a job offer or a prize, the relationships are what will get you there.

Now, there's a crucial element of this process that the steps above might not emphasize enough. In step six, the process ends; as we discussed in chapter 5, all brands eventually die. In general, the interest in our products tends to wax and wane like the bell-shaped curve in Figure 15-1. Marketers call this curve the Product Life-Cycle Curve.

Figure 15-1 illustrates the rise and fall in popularity of a hypothetical scientific product called "X." Maybe X is one of your signature research ideas. At first, the idea spreads slowly; your colleagues are skeptical. Then it gains momentum; your colleagues start calling you to ask about it. Then the market saturates and the curve tops out. Your colleagues start looking for alternative products that are more useful to them. Finally, X becomes part of science history—with a large role, a minor role, or most

The Product Life–Cycle Curve in Science

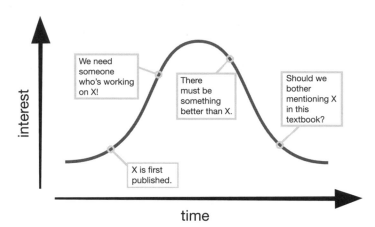

Figure 15-1   How interest in our scientific products waxes and wanes.
Source: Pete Yezukevich.

often, barely any role at all. Even if your product is destined to be in every science textbook, it will one day no longer be fashionable to do research on—at least, not among the same group of scientists who worked on it originally. Sucessful scientists understand this humbling truth, and they keep a small portfolio of intellectual products in various stages of development at once.

Now let's put these ideas together: the two parallel processes and the finite lifetimes of our products. You start with just a few friends and a big idea. And at the end, the idea may no longer be interesting, but you have gained many connections, friendships, and long-term collaborators.

When I look at the process this way, it seems to me that though our products may be fascinating, consequential, and even practical, the relationships are the true goal. I asked environmental microbiologist and former NSF director Rita Colwell about the rewards of a career in science, and she shared some of the joy of this goal with me. "Now I have wonderful friends, for example, several congressmen and -women with whom I've worked. I'm no longer in the position of Director of NSF and they are no longer in theirs, but we are still good friends. It's greatly enriched my life." Maybe you will win a fancy job or grant or prize, and you can flaunt it for a while. But that's not really the point. Ideally, your science will help create a new community of people—people who bond with you and with each other over your intellectual product. In a way, building a new community like this is the biggest external reward a person can ever get.

When the Product Is You

Let's take another look at the product development cycle, and now let's say the product is you—yourself, a scientist. You might learn your trade (step 1) by going to graduate school and taking lots of classes. Then you will need to decide how to brand yourself (step 2). That's about deciding what topic to work on, whether you will focus on theory or experiment, and what other brands you will be associated with, what archetype(s) will you project, and so on.

With any luck, by the time you finish graduate school you will have become some kind of Purple Cow, and there will be a small group of people you have inspired with your passion (step 3). These interested people might include your advisor and maybe some other members of the faculty. You have already been giving out free samples of your labor (or at least, approximately free) by working for a mere graduate-student stipend, and you have been listening to feedback from your advisor and other mentors.

Now, as you are finishing graduate school, it is time to move on to steps 4–6. That means meeting more people and developing more collaborations (step 4). At some point you will need to develop collaborators who admire you enough to write potent letters of recommendation for you. And at the same time you will need to develop at least one SRI that is associated with your name. When the SRI starts taking off, it will probably be time to apply for a better job or apply for a grant (step 5).

Many young scientists get to step 3 or step 4 and then slow down or stop. They don't realize that they need to continue to grow their networks and move new customers down their marketing funnels. They don't develop a portfolio of effective brands and refine them. They don't notice when their products have become obsolete or when the market is saturated. They stop pushing themselves to think of new, remarkable ideas and creative ways to market them.

Sometimes you will hear that such-and-such a young scientist is "doing his PhD over and over again" and that that's supposed to be a bad thing. I think the Product Development Cycle shines some light on the meaning of this assessment. Indeed, if your PhD was typical in terms of impact, i.e., not exactly earth-shaking, it is usually best to move on or expand your portfolio. But, if you launched a fantastic SRI during your PhD, destined to engender an entire new field, you would be crazy not to stick with it. Sticking with the same topic as your PhD is not necessarily bad; what's dangerous is not understanding this cycle or not having the guts to change direction or branch out when you need to.

Leadership: Starting a Movement

The Product Development Cycle is not just a business tool. It's also a story line. It's the plot to a series of Hollywood movies: *Gandhi, Erin Brokovich, Milk, The Social Network, Exit Through the Gift Shop*, and many others. Each of these movies tells the story of a charismatic person—a leader—who started a movement. Fortunately, we do not all need to reach the level of Gandhi to succeed in science, but I think we can all benefit by thinking about how to be a leader.

Derek Sivers, the Internet music-business hero I introduced in chapter 10, gave a talk about leadership at a 2010 TED conference, and received a standing ovation.[2] This talk taught me something about leadership that I want to share. During the talk, Sivers showed an unembellished amateur video he found on YouTube, one that would probably otherwise have gone unnoticed. It shows a lawn full of what might be college kids sitting on the grass. Loud music plays. One shirtless guy stands alone in the corner, dancing vigorously, haphazardly and enthusiastically to the loud, distorted music. He looks like a lone nut, a goofball.

Then suddenly another guy joins the shirtless dancer, copying his goofy dance, cavorting about in the corner of the lawn. Sivers explains, "Being a first follower is an underappreciated form of leadership. The first follower transforms a lone nut into a leader."

Indeed, about 20 seconds later, there is a second follower; another guy joins in the goofy dance. Sivers narrates, "Make sure outsiders see more than just the leader. Everyone needs to see the followers, because new followers emulate followers—not the leader." A few seconds later there are 5 people, 10 people, then suddenly the whole park is up and dancing. Sivers says, "As more people jump in, it's no longer risky. If they were on the fence before, there's no reason not to join now. They won't be ridiculed, they won't stand out, and they will be part of the in-crowd, if they hurry."

The video vividly demonstrates what some people call the "herd mentality." At first, the movement is slow to attract interest. But when it starts to look like many people are getting up to dance, the movement reaches a kind of "tipping point" and all the kids feel compelled to join so they aren't left out. In exactly this way, when people are uncertain about whether or not to concur with your scientific idea, they look to the opinions of others for guidance. Even scientists want to feel like they fit in, and they tend to feel uncomfortable if they stray too far from the mean.

We scientists may claim to be rugged individualists and logical thinkers, immune to the herd mentality. But it's just not possible to have that clarity all the time about all topics; there are too many decisions to make in our lives not to relinquish some of them to the wisdom of the herd. And, as you know, our social nature is hard-wired and acts even when we don't realize it; when we feel like we are combatting it on one front, it sneaks up and grabs us from behind. An example of this phenomenon is that sometimes a meeting of scientists can resemble a crowd of people just watching each other to see what ideas and job candidates everyone else seems to be going for.

Fortunately, there is another message for scientists in Sivers's talk. How can you start down the path of Gandhi, Erin Brokovich, or Harvey Milk? How do you become the one everyone is watching at a scientific meeting? Sivers tells us how; it takes a first follower and a second follower, two people who are willing to publicly take risks with you. That's the crux of Sivers's message. It's a message that applies to many circumstances. For example, in the movies I mentioned, Gandhi's first follower is his wife, Brokovich's is her lawyer, and Milk's is his lover.

Sivers says that most of your followers don't follow you; they follow your followers. So everyone else needs to see your first and second followers following you. He says that you must treat your first and second followers as equals, not underlings, or they will not follow you for long.

For scientists, that means finding two people who are keenly interested in your work, and willing to display that interest publicly. Maybe they show it by standing around your poster or by showing one of your slides in their talks. Maybe they show it by announcing that they have made you a job offer. But those two people are the key to your success; you need them badly, and you must do whatever it takes to maintain their faith in you.

The bottom line is that leadership is not about bossing people around or addressing a flock of underlings through a megaphone. It's about having a small core group of people who believe in you to the point where they will take risks to support your cause. They can publicly show their commitment to your ideas, and they can speak positively about your work in the way that's most credible—behind your back. That's the way to create the magical "buzz" that attracts more followers. This fact about human nature is one reason why scientists need mentors, collaborators, students, and friends.

But I'm Preaching to the Choir!

One scientist friend told me that she had been struggling to market her work, but felt like she was just preaching to the choir: only reaching people who already believed in her. I've sometimes had the same feeling myself, so let me take a moment and try to address this potential fear.

We've already talked about two ways marketing can help spread our messages beyond the proverbial choir. First of all, when you launch a new brand, nobody—choir or otherwise—has much of a preconceived notion about it because they don't know what it is yet. A new brand, with a new name, gives you a clean slate and a chance to attract followers who might have been out of reach.

Second, as we've just been discussing, people who won't listen to you will still listen to their friends. And their friends might listen to your preaching. You can preach to the choir about the importance of spreading the word, and maybe they will help you—if there's something in it for them.

But I want to point out that sometimes preaching is not what's needed at all. Sometimes the best way to make a difference is to take the risk of being someone else's first follower. Maybe there is someone you know who has a great idea, but no followers. You can become a follower yourself, and just let people observe and join the movement if they wish. Ironically, as Derek Sivers said, sometimes the best way to lead is to follow.

Marketing Zen

Even with the tools in this book, sometimes it's hard to move people from where they already stand, unless they are already inclined to move.

You might say marketing is about releasing a landslide, not picking up a mountain with a forklift. Part of what that means is that if you set your heart on attaining any one specific career goal in science, whether it's a certain job or a certain grant or a certain prize, you are heading for disappointment. This situation is counterintuitive for people like us, trained all through school to set specific goals—like getting an A in a class—and to labor and fight till we meet them.

Maybe it's better not to set out specifically to obtain a specific goal, like winning a Nobel prize. But if you stay flexible, and let your goal be helping other people through your work, you can accomplish a lot. In my mind I keep a kind of dazzling watercolor picture that shows the whole wide world spinning around on dreams and fears and demands—people helping each other and being helped in return. Looking at the world this way helps me stay flexible and even find a peaceful state, a kind of marketing Zen. I don't try to force my ideas on other people; I wait till they offer me their permission. I take pleasure when people criticize my work; they are helping me craft my message. I relax and exhale.

This vision I subscribe to goes hand in hand with the innocent archetype. Sometimes it can be hard to find any innocent feelings when you are embroiled in the struggles of launching and maintaining a scientific career; it's an attitude seldom taught in graduate school. But I find an innocent outlook essential for my science career—and maybe you will find it helpful too.

Live Like You Were Dying

As I interviewed scientists and communicators in the process of writing this book, I asked them about their work, their hopes, their dreams, their fears. It's impossible to have so many deep conversations without hearing wise words about following your heart and doing what you love.

> "A scientist who loves their work can motivate an entire laboratory and provide a spark that helps everyone be more creative and successful."
>
> —John Marburger, former science advisor
> to George W. Bush

Some of my favorite words on this topic are in a country song by Tim Nichols and Craig Wiseman, recorded by Tim McGraw. It's called "Live Like You Were Dying." *Billboard* magazine named it their number-one country song of 2004—the year Tim McGraw's father died. The message of the song is that we only have a short time here on Earth, so we have to make every moment count, even if that means taking risks. Yes, the idea is clichéd. But I'm pretty sure it's clichéd because it's right. So live like you were dying; this is the last piece of marketing advice I want to run by you.

In 2004, Steve Jobs announced that he had been diagnosed with a cancerous tumor in his pancreas. Jobs resisted the idea of surgery at first and tried to combat the cancer with a special diet. But later that year he underwent major surgery to remove portions of his pancreas, stomach, intestine and gall bladder. Even inventors, entrepreneurs, and technical wizards must face the limitations of mortality.

Reporters speculated about Jobs's health, describing his appearance as "gaunt" and his behavior "listless." The Bloomberg corporate news service accidentally published his obituary. Jobs responded with characteristic bravado; at a media presentation he showed off his healthy blood pressure stats and quoted Mark Twain: "Reports of my death are greatly exaggerated."

But when Jobs gave a commencement address at Stanford University in 2005, he was a changed man. He told the crowd of spellbound graduates, "Almost everything, all external expectations, all pride, all fear of embarrassment or failure—these things just fall away in the face of death, leaving only what is truly important. Remembering that you are going to die is the best way I know to avoid the trap of thinking you have something to lose. You are already naked. There is no reason not to follow your heart."[3]

Learning about marketing has taken me in many directions that I never anticipated. This direction—contemplating our limited time on Earth—is certainly the strangest. But I don't mind. When I decided to become a scientist, I was following my heart, and I imagine you were, too. To those who pursue it, science will always offer a feeling of anticipation and promise, the feeling that amazing things can happen. And that's a feeling worth sharing.

Take-Home Marketing Tips
for Scientists

- Marketing is all about human relationships; start a conversation.
- E-mail and social networks are great, but calling people or meeting with them is better.
- When you communicate with people, use their names.
- Carry a prop; tell a story.
- Give good customer service.
- Everyone is wondering: What's In It For Me? (WIIFM).
- Coin new terms (like "laser," "dinosaur," "dark matter").
- If there's a team, there should be a logo.
- Be positive and enthusiastic, and don't play the orphan.
- Creating new research questions is as good for your career as answering old ones.
- Promote your Signature Research Idea and it will promote you; promote the idea, not yourself.
- Focus your research; become the go-to person in your subfield.
- Make Beautiful Butterfly, Family Portrait, and Jenny Craig figures.
- Proposals: To sell a blender, show a picture of a margarita.
- The first audience for your papers is your coauthors.
- Attending a conference: it's all about the follow-up.
- Make videos about your work and put them online.
- Find an excuse to run an electronic newsletter or other large active mailing list.
- The page that pops up when someone Googles your name is the front door to your business online.
- Make a list of talking points *and* a list of feeling points.
- Use "Citizen Science" to spread the good word about scientific thought.
- Focus on the neglected middle of science's marketing funnel; and make that funnel flow.

Notes

Introduction

1. Heather Dewar, "Gingrich Endorses Environmental Goals," *Washington Inquirer*, April 25, 1996, http://articles.philly.com/1996-04-25/news/25658984_1_national-environmental-policy-institute-environmental-goals-environmental-laws.
2. Louis Lavelle, "University Endowments: Worst Year Since Depression," *Bloomberg Businessweek*, January 28, 2010.
3. Jeffrey Mervis, Jocelyn Kaiser, and Eli Kintisch, "House Approves Flat 2011 Budget for Most Science Agencies," *Science Insider*, December 9, 2010.
4. See, for example, "Doctoral Degrees: The Disposable Academic," *The Economist*, December 16, 2010.
5. See http://majorityleader.gov/YouCut/.
6. Jon D. Miller, Eugenie C. Scott, and Shinji Okamoto, "Public Acceptance of Evolution," *Science* 313, August 11, 2006, 765–66.
7. Michael L. Bean, "The Gingrich That Saved the ESA", *Environmental Forum*, Jan/Feb 1999.

Chapter 2 Notes

1. Theodore Levitt, "Marketing Myopia," *Harvard Business Review*, September–October 1975.
2. See http://turningpointemarketing.com/.
3. Possibly a quick firing of *mirror neurons*.
4. "Botanical Gardens Look for New Lures," *New York Times*, July 26, 2010.

Chapter 3 Notes

1. Dirk Zeller, *Telephone Sales for Dummies* (Indiana: Wiley Publishing, 2008).
2. See www.youtube.com/watch?v=loxJ3FtCJJA.
3. Seth Godin, *All Marketers Are Liars* (Surrey, England: Portfolio Hardcover, 2009).
4. D. S. Hamermesh and A. M. Parker, "Beauty in the Classroom: Professors' Pulchritude and Putative Pedagogical Productivity," NBER Working Paper No. W9853 (2003).
5. Dana Goodyear, "The Truffle Kid," *The New Yorker*, August 16, 2010.
6. "Advertising: Trying Harder," *Time Magazine*, July 24, 1964.
7. Al Ries and Jack Trout, *Positioning: The Battle for Your Mind* (USA: McGraw-Hill, 2000).
8. Anthony Bennett, ed., *The Big Book of Marketing* (USA: McGraw-Hill, 2010), 168.
9. Jeffrey Steingarten, *The Man Who Ate Everything* (New York: Vintage: 1998).
10. Nancy Clark, "An Interview with Madeline Albright," *WomensMedia*, June 15, 2009.
11. Christopher Beam, "Code Black: Of course Obama talks differently to different groups. So do most politicians." *Slate*, January 11, 2010, www.slate.com/id/2241114/.
12. Obama's dance on *The Ellen Degeneres Show*: see www.youtube.com/watch?v=RsWpvkLCvu4.
13. E.g., Joe Navarro, *Louder Than Words: Taking Your Career from Average to Exceptional with the Hidden Power of Nonverbal Intelligence* (New York: HarperCollins, 2010).

14. Michael Krantz, David S. Jackson, Janice Maloney, and Cathy Booth, "Apple and Pixar: Steve's Two Jobs," *Time Magazine*, October 18, 1999.

15. Carmine Gallo, *The Presentation Secrets of Steve Jobs: How to Be Insanely Great in Front of Any Audience* (USA: McGraw-Hill, 2009).

16. Hara Estroff Marano, "Marriage Math," *Psychology Today*, March 16, 2004.

Chapter 4 Notes

1. See www.cluetrain.com.

2. See www.tcho.com.

3. Glenn Collins, "Hoping Chefs Will Melt for Tcho Chocolate," *New York Times*, November 2, 2010.

4. Victoria McGovern, "The One-Minute Talk," *Science Careers*, March 13, 2009, http://sciencecareers.sciencemag.org/career_magazine/previous_issues/articles/2009_03_13/caredit.a0900034.

5. Seth Godin, *Permission Marketing: Turning Strangers into Friends and Friends into Customers* (New York: Simon & Schuster, 1999).

6. Lorna Flowers, Graham Rodgers, and Karen Keeley are Britons living and writing songs in Nashville.

7. J. L. Freedman and S. C. Fraser, "Compliance without pressure: The foot-in-the-door technique," *Journal of Personality and Social Psychology* 4 (1966): 195–202.

8. Benjamin Franklin, *The Autobiography of Benjamin Franklin* (New York: Dover Publications, 1996).

9. Greg Behrendt and Liz Tuccillo, *He's Just Not That Into You: The No-Excuses Truth to Understanding Guys* (New York: Simon and Schuster, 2004).

10. See, for example, Leil Lowndes, *How to Talk to Anyone: 92 Little Tricks for Big Success in Relationships* (USA: McGraw-Hill, 2003).

11. E.g., the Menger Sponge, Hausdorff dimension log 20 / log 3. Of course, science is not a true fractal; it has an inner scale of roughly 1 mind.

Chapter 5 Notes

1. Marty Neumeier, *The Brand Gap: How to Bridge the Distance Between Business Strategy and Design* (Berkeley: Peachpit Press, 2005), 2d ed.

2. See, for example, www.theladders.com/career-advice/10-ways-wreck-personal-brand.

3. There's a fun list of people known as the Father or Mother of something at www.wordiq.com/definition/List_of_people_known_as_the_father_or_mother_of_something.

4. For example, see Peter Woit, *Not Even Wrong: The Failure of String Theory and the Search for Unity in Physical Law* (New York: Basic Books, 2006).

5. Al Ries and Jack Trout, *The 22 Immutable Laws of Marketing: Violate Them at Your Own Risk!* (New York: HarperBusiness, 1994).

6. It was Bert Hinkler, an Australian aviator who accomplished the feat in 1931.

7. Gáspár Bakos et al., "HAT-P-11b: A Super-Neptune Planet Transiting a Bright K Star in the Kepler Field," *Astrophysical Journal* 710 (2010): 1724.

8. This more-or-less random sample has no female names. Are we failing to teach female science students to coin words, or do we ignore women scientists when they try to do so? I have a suspicion that this marketing issue might somehow lie near the roots of gender inequality in science.

9. There will be band-limited masks used on the James Webb Space Telescope, the

gargantuan, knock-your-socks-off, astronomer's dream-come-true that's intended to replace the Hubble Space Telescope later in this decade.

10. Marc Kuchner and Wesley Traub, "A Coronagraph with a Band-Limited Mask for Finding Terrestrial Planets," *Astrophysical Journal* 570 (2002): 900.

11. My coauthors included my second postdoctoral advisor, David Spergel, and the optics gurus Justin Crepp and Jian Ge.

12. Joshua Winn, Gregory Henry, Guillermo Torres, and Matthew Holman, "Five New Transits of the Super-Neptune HD 149026b," *The Astrophysical Journal* 675: 1531.

13. Sean Carroll, "How to Get Tenure at a Major Research University," Cosmic Variance Blog, *Discover* magazine, March 30, 2011, http://blogs.discovermagazine.com/cosmicvariance/2011/03/30/how-to-get-tenure-at-a-major-research-university/.

14. Chip Heath and Dan Heath, *Made to Stick: Why Some Ideas Survive and Others Die* (New York: Random House, 2007).

15. Ibid.

16. Seth Godin, *Purple Cow* (New York: Portfolio, 2002). The term comes from this ditty by Gelett Burgess: "I never saw a purple cow / I never hope to see one / But I can tell you anyhow / I'd rather see than be one" (first published in *The Lark*, May 1, 1895). Incidentally, Burgess is also credited with coining the word "blurb."

17. Michelle Nijhuis, "Prodigal Dogs," *High Country News*, February 2010.

18. The Physics arXiv Blog, April 30, 2010, www.technologyreview.com/blog/arxiv/25126/.

Chapter 6 Notes

1. Marc Kuchner, "Archetypes and Country Music," *Music Row Magazine*, August 2009.

2. *Cosmos: A Personal Voyage*, Episode 1: The Shores of the Cosmic Ocean, first broadcast by the Public Broadcasting System on September 28, 1980.

3. Alfred Kinsey's work spoke strongly of the lover archetype; he struggled with the burden of this characterization.

4. Lois P. Frankel, *Nice Girls Don't Get the Corner Office: 101 Unconscious Mistakes Women Make That Sabotage Their Careers* (New York: Business Plus, 2004), 80.

5. Carol S. Pearson, *The Hero Within: The Six Archetypes We Live By* (New York: Harper Collins, 1986)

Chapter 7 Notes

1. "Burrito Buzz—and So Few Ads," *Bloomberg Businessweek*, March 12, 2007, www.businessweek.com/magazine/content/07_11/b4025088.htm.

2. Cornelia Dean, *Am I Making Myself Clear? A Scientist's Guide to Talking to the Public* (Cambridge, MA: Harvard University Press, 2009), 52.

3. Jim Kukral, "How to Manipulate the Media & Get Free Publicity with Creative Marketing Ideas," March 5, 2007, www.jimkukral.com/how-to-manipulate-the-media-get-free-publicity-with-creative-marketing-ideas/.

4. Frank James, "Obama Kept Many Campaign Promises but Now Faces GOP Wall," National Public Radio, *Morning Edition*, January 18, 2011.

5. Maya Angelou, interview with Oprah Winfrey in honor of Angelou's 74th birthday on *The Oprah Winfrey Show*, first broadcast on April 4, 2004.

6. Nancy Baron, *Escape from the Ivory Tower: A Guide to Making Your Science Matter* (Washington, DC: Island Press, 2010), 189.

Chapter 8 Notes

1. E. Jerome McCarthy, *Basic Marketing: A Managerial Approach* (Homewood, Il: Irwin, 1960).
2. See, for example, John Seabrook, "The Price of the Ticket," *The New Yorker*, August 10, 2009.
3. J. Paul Peter and Jerry C. Olson, "Is Science Marketing?" *Journal of Marketing* Fall 1983, 111–25.
4. Nancy Baron, *Escape from the Ivory Tower: A Guide to Making Your Science Matter* (Washington DC: Island Press, 2010).
5. Sean Carroll, "How to Get Tenure at a Major Research University," Cosmic Variance Blog, *Discover* magazine, March 30, 2011, blogs.discovermagazine.com/cosmicvariance/2011/03/30/how-to-get-tenure-at-a-major-research-university/.

Chapter 10 Notes

1. "Cisco Visual Networking Index: Forecast and Methodology, 2009–2014," Cisco, June 2, 2010, www.cisco.com/en/US/solutions/collateral/ns341/ns525/ns537/ns705/ns827/white_paper_c11-481360_ns827_Networking_Solutions_White_Paper.html.
2. Peter S. Fiske, *Put Your Science to Work: The Take-Charge Career Guide for Scientists* (American Geophysical Union, 2000).
3. Peter S. Fiske, *Opportunities: Career Advantages of Collaboration* January 9, 2009, http://sciencecareers.sciencemag.org/career_magazine.
4. Kathy A. Svitil, "The 50 Most Important Women In Science," *Discover*, November 2002, discovermagazine.com/2002/nov/feat50/.
5. Imke de Pater, Heidi B. Hammel, Seran G. Gibbard, and Mark R. Showalter, "New Dust Belts of Uranus: One Ring, Two Ring, Red Ring, Blue Ring," *Science* 312, (2006): 92–94.
6. Itay Sagi and Eldad Yechaim, "Amusing titles in scientific journals and article citation," *Journal of Information Science* 34 (October 2008): 680–87.
7. Edward R. Tufte, *The Visual Display of Quantitative Information*, 2d ed. (Cheshire, CT: Graphics Press, 2001); also Edward R. Tufte, *Envisioning Information* (Cheshire, CT: Graphics Press, 1990).
8. Marcie Sillman, "CD Baby Finds Success in Online Music Niche," National Public Radio, *Morning Edition*, December 28, 2004.
9. Derek Sivers, "Attending a music biz conference? Here's the REAL way to do it," http://sivers.org/conferences.

Chapter 11 Notes

1. See www.nsaspeaker.org.
2. See, for example, Malcolm Gladwell, *Blink: The Power of Thinking Without Thinking* (New York: Little, Brown & Company, 2005).
3. Jhn Medina, *Brain Rules: 12 Principles for Surviving and Thriving at Work, Home, and School* (Seattle: Pear Press, 2008).
4. Syd Field, *Screenplay: The Foundations of Screenwriting; A Step-by-Step Guide from Concept to Finished Script* (New York: Dell, 1979); Randy Olson, *Don't Be Such a Scientist* (Washington, DC: Island Press, 2009).
5. Joseph Campbell, *The Hero with a Thousand Faces* (Princeton, NJ: Princeton University Press, 1949).

Chapter 12 Notes

1. Alexis Madrigal, "Wired Science Scores Exclusive Twitter Interview with the Phoenix Mars Lander," *Wired Science*, May 30, 2008.
2. Mark Pfeifle, "A Nobel Peace Prize for Twitter?" *Christian Science Monitor*, July 6, 2009.
3. See www.youtube.com/watch?v=j50ZssEojtM.
4. "Too Many Blogs?" www.webdesignerdepot.com/2011/01/too-many-blogs/.
5. See www.usa.gov/Topics/Reference_Shelf/News/Blog/science.shtml.
6. See www.chrisbrogan.com/how-to-manage-twitter/.
7. Nick O'Neil, "Facebook Serves as Good Predictor of Election Results," AllFacebook. com, November 3, 2010, www.allfacebook.com/facebook-serves-as-good-predictor-of-election-results-2010-11. For the U.S. Senate races, the figure was 81 percent.
8. James Carville and Paul Begala, *Buck Up, Suck Up . . . and Come Back When You Foul Up: 12 Winning Secrets from the War Room* (New York: Simon and Schuster, 2002), 88.
9. See http://network.nature.com/.
10. Virginia Gewin, "Social media: Self-reflection, online," Nature Jobs, *Nature* 471 (2011): 667–69, www.nature.com/naturejobs/2011/110331/full/nj7340-667a.html.

Chapter 13 Notes

1. "It's True: Spaghetti Tacos 'Expert,' Prof. Robert Thompson, Has Now Been Interviewed by 78 Different NYT Reporters," *The NYTpicker*, October 5, 2010, www.nytpick.com/2010/10/its-true-spaghetti-tacos-expert-prof.html.
2. Geoff W. Beattie, *Get the Edge: How Simple Changes Will Transform Your Life* (Terra Alta, WV: Headline Book Publishing, 2011).
3. A *factlet* is a little fact—not to be confused with a *factoid*, which is a statement that resembles fact, but is not, in fact, a fact.
4. Richard Boleslavsky, *Acting: The First Six Lessons* (New York: Routledge, 1933).
5. Alan Alda, reported by Daniel Grushkin, "Try Acting Like a Scientist," *The Scientist*, August 5, 2010.
6. Cristina Eisenberg, *The Wolf's Tooth: Keystone Predators, Trophic Cascades, and Biodiversity* (Washington, DC: Island Press, 2010).

Chapter 14 Notes

1. Delaware Republican senatorial candidate Christine O'Donnell on *The O'Reilly Factor*, Fox News, November 16, 2007.
2. *Rising Above the Gathering Storm, Revisited: Rapidly Approaching Category 5* (Washington, DC: National Academies Press, 2010).
3. National Science Foundation, Division of Science Resources Statistics, *Scientists, Engineers, and Technicians in the United States: 2001*, NSF 05-313, Richard E. Morrison, Project Officer, with Maurya M. Green (Arlington, VA: National Science Foundation, 2005).
4. Everett M. Rogers, *Diffusion of Innovations*, 5th ed. (New York: Free Press, 2003).
5. See www.scientificamerican.com/page.cfm?section=calling-all-scientists.
6. Regina Corso, *The Harris Poll* no. 86, August 4, 2009, www.harrisinteractive.com/vault/Harris-Interactive-Poll-Research-Pres-Occupations-2009-08.pdf.
7. Alan Leshner, "Life Science Contributions to the 21st Century," a lecture given at Preserving National Security: The Growing Role of the Life Sciences, a conference convened by the Center for Biosecurity of the University of Pittsburgh Medical Center, March 3, 2011, www.upmc-biosecurity.org/website/events/201103-lifesci/videos.html.

8.   See www.worldwildlife.org/species/finder/polarbear/geoffyork.html.

9.   "Schools should not limit origins-of-life discussions to evolution, Republican Legislators Say," *Stillwater Gazette*, September 27, 2005.

10. See http://boinc.berkeley.edu/.

11. Paul Rincon, "Teacher Finds New Cosmic Object," *BBC News*, August 5, 2008.

Chapter 15 Notes

1.   See, for example, Steven Silbiger, *The Ten-Day MBA: A Step-By-Step Guide to Mastering the Skills Taught in America's Top Business Schools* (New York: William Morrow, 1993).

2.   See http://sivers.org/ff.

3.   Steve Jobs, Stanford University Commencement Speech, 2005; see www.youtube.com/watch?v=D1R-jKKp3NA.

Further Reading

 Note from the Author: Marketing seems to be one of those subjects that takes a lifetime to learn. Maybe that's because once you get good at marketing your work to one group of people, you often want to move on and meet another group of people with different needs and a different culture. Or maybe it's because the tools and technologies we have for connecting with people keep changing at the pace of an angry neutrino. In any case, the books in the list that follows can help you take your next step—in any of several different directions.

When I am reading marketing books, one thing I like to look for is how the book itself is marketed. For example, you will notice that many marketing books have provocative, memorable titles. You'll often find that they are written using anecdotes that entertain you and pique your curiosity. I tried using some of the same tricks to write this marketing book.

And this here book/project is not done, y'all. I need your help to finish it. If you know a great science marketing trick that's not mentioned in the book, please share it with me and with our colleagues by posting it to the Facebook group "Marketing for Scientists." You can find a link to this group at www.marketingfor scientists.com. Or start your own movement and change the world.

Thanks for reading this book and thinking about it.

Salesmanship and Relationship Building

Permission Marketing: Turning Strangers into Friends and Friends into Customers
Seth Godin
New York: Simon & Schuster, 1999

> Godin makes an analogy between relationship building in business and the process of courting a lover, pushing the analogy as far as he can. He even calls the marketing funnel the "funnel of love." When I'm reading Godin's books, I like to keep in mind his concept for the purpose of a book in today's world: a book is a "souvenir of an idea." In other words, expect to find a short book that's tighly focused on one idea, written with fun in mind.

All Marketers Are Liars
Seth Godin
Surrey, England: Portfolio Hardcover, 2009

> The irony of the title is that it's not really the marketers who are liars, but rather that we, the customers, lie to ourselves all the time. But without these lies, life would be sad. It's a Faustian bargain that everyone makes, even scientists.

Never Eat Alone: And Other Secrets to Success, One Relationship at a Time
Keith Ferrazzi with Tahl Raz
New York: Doubleday, 2005

> This book drove home for me the difference between networking and relationship building, and helped me understand the business mindset. It suggests this charmingly honest way to break away from a conversation at a meeting. "There are so many wonderful people here tonight, I'd feel remiss if I didn't try to get to know a few more of them."

Yes! 50 Scientifically Proven Ways to Be Persuasive
Noah J. Goldstein, Steve J. Martin, and Robert B. Cialdini
New York: Free Press, 2008

> Fun tidbits that illustrate, among other things, how adults behave like gullible children when their attention is divided. Each way in which to be persuasive points to a human weakness or two—and on this list I found many faults of my own.

Nice Girls Don't Get the Corner Office: 101 Unconscious Mistakes Women Make That Sabotage Their Careers
Lois P. Frankel
New York: Business Plus, 2004

> A general, no-nonsense book on sales, time management, and nonverbal communication techniques for the workplace; in fact, only a few of the tips are exclusively for women.

Louder Than Words: Taking Your Career from Average to Exceptional with the Hidden Power of Nonverbal Intelligence
Joe Navarro
New York: HarperCollins, 2010

> An ex-FBI agent provides advice on how to read people and make a good impression. Some of it is obvious, but some of it isn't. Juries find men in two-button suits more honest than men in three-button suits. When you're standing talking to someone at a meeting, it's best to place your feet at a 45-degree angle from the other person's feet.

Blink: The Power of Thinking Without Thinking
Malcolm Gladwell
New York: Little, Brown & Company, 2005

> Yes, you are trained to be a logical thinker, but it takes a few hundred milliseconds for that training to kick in. This book helped open my eyes to the importance of those milliseconds when first impressions are made—crucial moments for marketers. Thinking without thinking is powerful all right, but don't mistake this book for a manual on how to develop a Jedi-like mastery of some hidden strength; the power of preconscious thinking is more like the power of a thunderstorm, a force to be admired and feared.

Wikinomics: How Mass Collaboration Changes Everything
Don Tapscott and Anthony D. Williams

With Web 2.0 tools like Facebook and wikis, it has suddenly become easy to harness vast amounts of brainpower to solve whatever problem you're working on in a blink of an eye. But this book's not just about harnessing the brains of others. If you think about it the right way, this is a book about relationship building, too. I've already bought copies of this book for five or six friends.

BRANDING AND ARCHETYPES

Positioning: The Battle for Your Mind
Al Ries and Jack Trout
USA: McGraw-Hill, 2000

A classic book on a key marketing concept. I found it somewhat incoherent in parts, but still full of fun stories à la *Mad Men*. It has tales of Cadillac, Milk Duds, Tylenol, Exxon, Xerox, Cutty Sark, IBM, Goodyear, Scope, and many more brands that have shaped our American culture.

The 22 Immutable Laws of Marketing: Violate Them at Your Own Risk!
Al Ries and Jack Trout
New York: HarperBusiness, 1994

Another classic. I summarized what I think are the most relevant parts in chapter 5, but it might be fun to read anyway for the stories.

Purple Cow
Seth Godin
New York: Portfolio, 2002

In this little book, Godin explains why in today's marketplace it's good to be a little crazy (passionate about something), and why it's dangerous to not take risks. I think scientists understand this truth intuitively, but we sometimes lose sight of it during graduate school.

Made to Stick: Why Some Ideas Survive and Others Die
Chip Heath and Dan Heath

Ideas that will help make your messages compelling and infectious. Many of these ideas apply to science, though some must be used with caution. For example, the book points out that a good way to make your messages memorable is to put them in rhyme. That works on Twitter, but it's probably not going to fly in *JAMA*.

The Hero and the Outlaw: Building Extraordinary Brands Through the Power of Archetypes
Margaret Mark and Carol S. Pearson

A fun and frighteningly insightful book about your favorite brands and why we love them. This book explains the system of twelve archetypes I adapted in chapter 6.

The Hero Within: Six Archetypes We Live By
Carol S. Pearson

> This isn't a marketing book; it's self-help popular psychology book. But it shows another facet of the far-reaching theory of archetypes. It can be good medicine for days when you feel like the orphan.

ACADEMIC LIFE

A PhD Is Not Enough: A Guide to Survival in Science
Peter J. Feibelman
New York: Perseus Books, 1993

> The book that first shattered my notion that I could rely on graduate education alone to bring me the scientific career I wanted. Though many experts have responded to the call of the book's title over the last two decades, the problems with academia described herein are still rampant, and the remedies the book offers still help.

The Shortcut to Persuasive Presentations
Larry Tracy
North Charleston, SC: Imprint Books, 2003

> The former head of the Pentagon's top briefing team explains his approach to giving talks. I especially like the tips on how to handle aggressive audiences.

The Visual Display of Quantitative Information
Edward Tufte
Cheshire, CT: Graphics Press, 1992

> A spectacular tome filled with age-old wisdom and modern tips on how to make intuitive charts and graphs. Perfect for the coffee table of your office when you get tenure.

Put Your Science to Work: The Take-Charge Career Guide for Scientists
Peter Fiske
Washington, DC: American Geophysical Union, 2000

> This book is the all-around career guide that *Marketing for Scientists* is not. It's especially useful if you think you might want to leave the traditional academic science career path. Fiske writes the Opportunities column on the AAAS website, Science Careers.org.

THE PUBLIC, THE GOVERNMENT, THE INTERNET, AND THE PRESS

The Cluetrain Manifesto
Rick Levine, Christopher Locke, Doc Searls, and David Weinberger
New York: Perseus Books, 2001

> "Markets are conversations." That's the first of 95 "theses" these authors put forth in 1999 on a revolutionary website, www.cluetrain.com, describing the effects of the newly popular Internet on the world marketplace. You can buy a hard copy or download it from the website, but whatever form you prefer, *Cluetrain* is a must read. Its

appearance marked the beginning of a new era for marketing, one where customer interaction and customer participation can no longer be considered accessories to a sucessful business model.

How to Master the Media
George Merlis
Los Angeles: JAAND Books, 2007
   A former executive producer of *CBS Morning News* explains the fundamentals of the "media training" approach to working with the press; this is an essential technique for scientists to know. Among other things, it will help you avoid the CUPP (Cover Up Private Parts) position, a defensive posture that looks bad on TV.

Don't Be Such a Scientist: Talking Substance in an Age of Style
Randy Olson
Washington, DC: Island Press 2009
   A tenured marine biologist, who quit and moved to Hollywood, explains what it's like to go to film school and make movies. This book is not just for scientists interested in acting and movie-making (which maybe should be all of us, at some level); it explains what the public thinks of scientists, why they think that way, and what we can do about it. My favorite passage is when Randy first shows up at film school and encounters the words written by the door: Reality Ends Here.

Am I Making Myself Clear? A Scientist's Guide to Talking to the Public
Cornelia Dean
Cambridge, MA: Harvard University Press, 2009
   Read this to find out what it's like to be a *New York Times* science writer, calling scientists and trying to put together a story. This book has tips on blogging, serving as an expert witness, and influencing policy as well.

Escape from the Ivory Tower: A Guide to Making Your Science Matter
Nancy Baron
Washington, DC: Island Press, 2010
   I've quoted several ideas from this book, which has handy sections called "Preconceptions policymakers have of scientists" and "Preconceptions scientists have of policymakers." It strikes me that understanding and applying these preconceptions is a great example of marketing for scientists (specifically, it's "positioning").

Thinking Like Your Editor: How to Write Great Serious Nonfiction--and Get It Published
Susan Rabiner and Alfred Fortunato
   My agent, Judy Heiblum, insisted that I read this book. If you told me you wanted to write a book, I'd insist you read it, too. It covers writing a book proposal, the process of researching and writing the book, and crucially, how to get along with the various actors you'll meet in the literary business. And I love that it's got marketing built into the title: learning to think like someone you're not.

The Savvy Author's Guide to Book Publicity
Lissa Warren
New York: Carrol and Graf Publishers, 2004

> Several of my colleagues have asked me for advice about how to market a book. Since *Marketing for Scientists* is my first book, I've mostly told them: ask me again in a year! But I can pass along the advice my publisher gave me on this front. (1) Tweet. (2) Read *The Savvy Author's Guide to Book Publicity*. It's not entirely up to date on all the various ways you can promote a book online. But it contains some fantastic suggestions—like writing a quiz on your subject matter.

Marketing for Engineers, Scientists, and Technologists
Tony Curtis
West Sussex, England: John Wiley & Sons Ltd., 2008

> A textbook-style introduction to marketing, and a good book to read next if you are in this game for the money. That is, literally going into business, selling something you invented, or getting an MBA.

Index

About Island Press

Since 1984, the nonprofit Island Press has been stimulating, shaping, and communicating the ideas that are essential for solving environmental problems worldwide. With more than 800 titles in print and some 40 new releases each year, we are the nation's leading publisher on environmental issues. We identify innovative thinkers and emerging trends in the environmental field. We work with world-renowned experts and authors to develop cross-disciplinary solutions to environmental challenges.

Island Press designs and implements coordinated book publication campaigns in order to communicate our critical messages in print, in person, and online using the latest technologies, programs, and the media. Our goal: to reach targeted audiences—scientists, policymakers, environmental advocates, the media, and concerned citizens—who can and will take action to protect the plants and animals that enrich our world, the ecosystems we need to survive, the water we drink, and the air we breathe.

Island Press gratefully acknowledges the support of its work by the Agua Fund, Inc., The Margaret A. Cargill Foundation, Betsy and Jesse Fink Foundation, The William and Flora Hewlett Foundation, The Kresge Foundation, The Forrest and Frances Lattner Foundation, The Andrew W. Mellon Foundation, The Curtis and Edith Munson Foundation, The Overbrook Foundation, The David and Lucile Packard Foundation, The Summit Foundation, Trust for Architectural Easements, The Winslow Foundation, and other generous donors.

The opinions expressed in this book are those of the author(s) and do not necessarily reflect the views of our donors.